Revision of the *Modified Mouthparts* Species Group of Hawaiian *Drosophila* (Diptera: Drosophilidae)

Revision of the *Modified Mouthparts* Species Group of Hawaiian *Drosophila* (Diptera: Drosophilidae)

The *Ceratostoma*, *Freycinetiae*, *Semifuscata*, and *Setiger* Subgroups, and Unplaced Species

Karl N. Magnacca and Patrick M. O'Grady

UNIVERSITY OF CALIFORNIA PRESS
Berkeley • Los Angeles • London

University of California Press, one of the most distinguished university presses in the United States, enriches lives around the world by advancing scholarship in the humanities, social sciences, and natural sciences. Its activities are supported by the UC Press Foundation and by philanthropic contributions from individuals and institutions. For more information, visit www.ucpress.edu.

University of California Publications in Entomology, Volume 130

University of California Press
Berkeley and Los Angeles, California

University of California Press, Ltd.
London, England

Printed in the United States of America

Cataloging-in-Publication data for this title is on file with the Library of Congress.

ISBN 978-0-520-09873-2 (pbk. : alk. paper)

The paper used in this publication meets the minimum requirements of
ANSI/NISO Z39.48-1992 (R 1997) (Permanence of Paper).

CONTENTS

ACKNOWLEDGEMENTS

Thanks to Betsy Gagné of the Hawai'i State Division of Forestry and Wildlife, Ed Misaki and Stefanie Loo of The Nature Conservancy of Hawai'i, and Tim Tunison and Rhonda Loh of Hawai'i Volcanoes National Park for their assistance with collecting permits; Neal Evenhuis, Keith Arakaki, and Shepherd Myers of the B. P. Bishop Museum, Dan Rubinoff, Will Haines, and Luc LeBlanc of the University of Hawai'i–Mānoa, and Kim Goodger of the Natural History Museum, London for their support with examining and borrowing specimens from their respective collections; David Foote of USGS–PIERC Kīlauea Field Station for allowing us to include specimens in this revision; Phil Ward of the University of California–Davis and two anonymous reviewers for their comments on this manuscript; and Steve Montgomery for his tireless collecting, without which many of these species would remain unknown.

ABSTRACT

The *modified mouthparts* group is perhaps the largest of the four major Hawaiian *Drosophila* clades, yet has received relatively little taxonomic attention during the past 40 years. This paper reviews several of the smaller subgroups, with descriptions of 22 new species: *D. orascopa* n. sp. and *D. wikani* n. sp. in the *ceratostoma* subgroup; *D. anapuu* n. sp., *D. dentilabia* n. sp., *D. kualii* n. sp., and *D. nalomano* n. sp. in the *freycinetiae* subgroup; *D. mandibulata* n. sp., *D. peloristoma* n. sp., and *D. wawae* n. sp. in the *semifuscata* subgroup; *D. desallei* n. sp. in the *setiger* subgroup; and 12 unplaced species, *D. apiki* n. sp., *D. barbata* n. sp., *D. gladius* n. sp., *D. incongruens* n. sp., *D. komohana* n. sp., *D. lelolua* n. sp., *D. omnivora* n. sp., *D. toxacantha* n. sp., *D. umiumi* n. sp., *D. wahihuna* n. sp., *D. waikamoi* n. sp., and *D. wikstroemiae* n. sp. The male of *D. acrostichalis* Hardy, 1965 is described for the first time, and *D. apoxyloma* Hardy, 1965 is reduced to a new junior synonym of *D. setiger* Grimshaw, 1901.

INTRODUCTION

The *modified mouthparts* species are a large, heterogeneous group, the males of which possess various labellar characters used during courtship and mating (Spieth 1966). These range from relatively modest changes to the labellar setae to thickened, tusk-like setae and ornate projections of the labellar sclerites. Over the course of the last 40 years, the *modified mouthparts* group has been the least studied group of Hawaiian *Drosophila*, with only seven new species (Hardy 1966; Hardy and Kaneshiro 1968; Hardy and Kaneshiro 1969; Hardy and Kaneshiro 1975a; Takada and Yoon 1989) described between D. E. Hardy's Insects of Hawai'i monograph (1965) and 2003 when the *mimica* subgroup was revised (O'Grady et al. 2003). Somewhat surprisingly, several of the earlier descriptions (Bryan 1934; Bryan 1938; Grimshaw 1901) made no mention of the bizarre development of the labellum, despite their detailed descriptions of other characters. Even later (Hardy 1965), characters such as wing coloration were treated as more indicative of species relationships than labellar characters. No doubt this was due in part to many subgroups being represented by only one or a few species, thus appearing to be an odd outlier rather than a representative of a larger lineage.

Recently, Magnacca and O'Grady (2007) proposed a subgroup classification scheme to better organize the diversity of the *modified mouthparts* group and prepare for future revisionary work. There are currently 83 described species in this group; an additional 22 new species are described here and one (*D. sadleria*) is newly-recognized as a member, for a total of 106 species. At least another 30 taxa have been collected, but await description. The rather haphazard collecting of *modified mouthparts* species to date, compared to the intensive surveys done for the *picture wing* group, means that many more likely remain to be discovered. The classification scheme of Magnacca and O'Grady (2007) is modified slightly to accommodate the additional new species, including reduction of *D. adventitia* to an unplaced species rather than a monotypic subgroup, and splitting up the *bridwelli* subgroup, placing its members into the *nanella* and *semifuscata* subgroups. This paper provides descriptions for all species currently known from four of the twelve *modified mouthparts* subgroups, as well as "unplaced" *modified mouthparts* species (Table 1; see keys and discussions below). Keys to the species of the *fuscoamoeba*, *mitchelli*, *nanella*, and *scolostoma* subgroups (from which there are no undescribed species currently known) are also included.

Many of the unplaced species described here have somewhat generalized mouthparts that blur the differences between subgroups (e.g. *D. barbata* n. sp., *D. komohana* n. sp., *D. omnivora* n. sp., and *D. wikstroemiae* n. sp.), rather than the extreme forms that exist in previously-described species. It is expected that the placement of some species may be shifted if and when a thorough phylogenetic study is completed. However, most have not been collected in decades, and

Table 1. Subgroup classification, distribution, and rearing records for new species described in this paper.

Subgroup Species	Distribution	Rearing Records	Hosts
ceratostoma			
orascopa	Hawai'i		
wikani	Hawai'i, Maui?[1]		
freycinetiae			
anapuu	Hawai'i	1	*Freycinetia* stems
dentilabia	Maui, Lāna'i?[2]	1?	*Freycinetia* leaf axils?
kualii	Maui		
nalomano	O'ahu	2	*Wikstroemia*, *Touchardia* bark
semifuscata			
mandibulata	Maui, Moloka'i		
peloristoma	O'ahu	4	*Acacia*, *Myrsine*, *Nestegis* sap flux
wawae	Hawai'i		
setiger			
desallei	Hawai'i	1	*Pisonia* bark
unplaced			
apiki	Kaua'i	1	*Labordia* bark
barbata	Hawai'i	8	*Charpentiera*, *Clermontia*, *Freycinetia*, *Pisonia*, *Urera* bark & stems
gladius	Hawai'i		
incongruens	Maui		
komohana	Maui		
lelolua	Lāna'i	1	*Freycinetia* leaf axils
omnivora	O'ahu	4	fern frond, *Strongylodon* pod, *Touchardia* & *Urera* bark
toxacantha	Hawai'i	3	*Cheirodendron trigynum* bark
umiumi	Hawai'i	2	*Clermontia* bark
wahihuna	Kaua'i		
waikamoi	Maui		
wikstroemiae	O'ahu	1	*Wikstroemia* bark

[1] See Discussion under this species.

[2] Specimen from Lāna'i was reared; unpublished description matches *D. dentilabia* but is incomplete, and specimen was not available for comparison.

the likelihood of obtaining fresh specimens for morphological or DNA analysis is low without a renewed effort to rear flies from host plant material. Much of the difficulty with working on this group is the relative rarity of most species in collections. Most are not attracted to the typical banana and mushroom baits used by Hawaiian *Drosophila* workers, and are most readily obtained by more labor-intensive rearing from host substrates. Large series are available for relatively few species, meaning that intraspecific variability is very difficult to address.

This paper covers much of the morphological diversity of the *modified mouthparts* group, but considerable taxonomic work remains. The following three subgroups are not considered in this paper. The *hirtitarsus* subgroup has representatives on all islands which probably represent multiple island-endemic species, but at present cannot be reliably separated. In examining specimens it became clear that the O'ahu specimens called "*hirtitarsus*-like" are in fact *D. goureaui* (=*D. mycetophila* Hardy, 1965). This species was not previously recognized as belonging in the *modified mouthparts* group due to the relatively minor mouthpart modifications in this subgroup. It is readily separated from *D. hirtitarsus* by its all-yellow pleura. There are also numerous specimens from this subgroup from Hawai'i (referred to as "aberrations or possibly new species" in Hardy 1966), and a few from Kaua'i, in the University of Hawai'i Insect Museum (UHIM). These populations probably represent genetically isolated species, but no characters have been found to reliably distinguish them from typical *D. hirtitarsus* of Maui Nui.

The *quadrisetae* subgroup is somewhat more diverse, and appears to have multiple representatives on each island. However, these can be very difficult to distinguish, and like *D. hirtitarsus* there are morphologically identical forms on multiple islands. Such a situation is possible but rare in Hawaiian drosophilids – in fact, it appears to occur in several species of the *semifuscata* subgroup discussed below – and should be investigated further.

The *dissita* subgroup has a large number of undescribed species – at least 18 are known in collections, and probably many more remain uncollected – and its revision will be a major undertaking. The fact that five undescribed species from the island of Hawai'i have been reared but never otherwise been collected suggests that many more from other islands remain to be discovered. The diversity of mouthpart forms is much greater than that within other subgroups, but they appear to grade into each other. Furthermore, when the undescribed species are considered it becomes clear that the *dissita* and *mimica* subgroups cannot be reliably separated as currently constituted, and each appears to be paraphyletic with respect to portions of the other. The *quadrisetae* subgroup may also be derived from within the *dissita-mimica* complex. It is likely that rearing, even on a relatively small scale, will result in a major increase in our understanding of the diversity of *modified mouthparts* species.

MATERIALS AND METHODS

Material comes from historic pinned specimens, recent collections (1997–2007, both pinned and in ethanol), and flies reared from hosts. Most specimens came from either sweeping vegetation or rearing from hosts, as these species are not attracted to standard banana/mushroom baits. Historic material is from the University of Hawai'i Insect Museum (UHIM), B. P. Bishop Museum (BPBM), and the Natural History Museum, London (BMNH). Holotypes and allotypes of new species have been deposited at the BPBM, and paratypes at BPBM, UHIM, and the Hawai'i Volcanoes National Park Insect Collection (HVIC) as noted. Additional material examined is in the research collection of the O'Grady lab at the University of California, Berkeley (UCPO).

Measurements were taken with an eyepiece micrometer. Measurements are the same as those used by Magnacca and O'Grady (2008), and are defined as follows. TL = thorax length, distance from the anterior notal margin to the posterior apex of the scutellum. WL = wing length, maximum distance from the humeral crossvein to the apex of the wing. TL/WL = ratio of thorax length to wing length. HW = head width, greatest distance across the eyes. HW/FS = ratio of head width to width between the eyes above the frontal suture. HW/TL = ratio of head width to thorax length. CI = costal index, length of costa from subcostal break to R_{2+3}/length of costa from R_{2+3} to R_{4+5}. 4V = fourth vein index, length of M from crossvein dm–cu to apex/length of M from crossvein r-m to crossvein dm–cu. 5X = length of CuA_1 from crossvein dm–cu to apex/length of crossvein dm–cu. 4C = length of costa from R_{2+3} to R_{4+5}/length of M from crossvein r–m to crossvein dm–cu. M = length of CuA_1 from crossvein dm–cu to apex/length of M from crossvein r–m to crossvein dm–cu.

Abbreviations for collectors are JWB (John W. Beardsley), HLC (Hampton L. Carson), DEH (D. Elmo Hardy), KH (Kirsten Heckmann), WBH (William B. Heed), KYK (Kenneth Y. Kaneshiro), KNM (Karl N. Magnacca), SLM (Steven L. Montgomery), JPM (John P. Murphy), PMO (Patrick M. O'Grady), ATO (Alan T. Ohta), RCLP (Robert C. L. Perkins), WDP (William D. Perreira), HTS (Herman T. Spieth), OHS (Otto H. Swezey), LHT (Lynn H. Throckmorton), and FXW (Francis X. Williams).

Mouthpart and genitalia illustrations were made from dissected and cleared specimens, except as noted. See Fig. 1 for terminology. Membranous areas are indicated with light gray, and brown spinose setae with dark gray. Stippling in genitalia illustrations indicates microtrichose regions. Sclerotization of the labellum is variable, and the boundaries of sclerites may appear more diffuse in some specimens than in illustrations. In many species – most notably members of the *nanella* and *semifuscata* subgroups, and most of the unplaced species – a pair of dorsoapical sclerites, here referred to as the preoral sclerites, are enlarged and conspicuous. The proboscis is shown from the left as it would appear fully extended ventrally, with the apex of the labellar sclerite down; because the labellum is derived from the labial palp, this means that anatomically dorsal is to the left. The proboscis will appear narrower in dry specimens due to contraction of the membranous area between the labium and preoral sclerite. Setae are counted from the dorsal end of the labellum. The terms "round" and "flattened" in reference to the labellar setae are used to describe the cross-sectional shape. For examples of the mouthparts of subgroups and species not illustrated here, see Magnacca and O'Grady (2007).

Leg illustrations show all setae and setulae. Unlike most other Hawaiian *Drosophila* groups, there is not a sharp differentiation between the elongate cilia of the male front legs used in courtship and the short, straight setulae similar to those found on the other legs. In the genitalia, the parameres, surstyli, and caudal portion of the hypandrium are typically at least partially concealed and/or variable in appearance depending on the precise angle of view. Therefore, their shape as shown in the illustrations should not be taken as reliable characters for species separation.

The age of specimens in collections can be a major impediment to separation of closely-related *Drosophila* species. Coloration can be a reliable and consistent character in life, but pinned specimens often discolor over time, seriously reducing its utility for older material. Specimens in ethanol may retain their color better, but will bleach after several years in denatured alcohol or methylated spirits. Most species are described from relatively old (>30 years) pinned specimens, which are often discolored. This is especially the case for the legs, pleura, and abdomen, where the cuticle is thin. Light-colored areas may appear darker than in life due to the underlying musculature showing through, while areas that are nearly black often fade to brown in pinned specimens. Therefore, when using the identification keys, caution should be used when interpreting coloration characters depending on the condition of specimens on hand and those used for the descriptions.

KEY TO SUBGROUPS AND UNPLACED SPECIES

BASED ON MALES ONLY

1. Mesonotum bright yellow with two longitudinal brown stripes just inside the dorsocentral rows; wing infuscated along the entire anterior margin and along vein CuA_1 (male unknown, but likely to key out here based on the distinctive female). Hawai'i .... *D. tetraspilota* Hardy

— Not marked as above .... 2.

2. (1) Labellum with only a fringe of relatively short, thin, hair-like setae and without strong spine-like setae (rarely with short peg-like setae), the setae distinct from those of non-*modified mouthpart* species only in being arranged in a row and arising from a weakly sclerotized plate (labella often tightly folded together in pinned specimens).. 3.

— Labellum distinctly modified, with at least one spine-like seta, a group of elongate setae, or a sclerotized appendage .... 5.

3. (2) Wings with the entire anterior margin infuscated, often curving around to the dm–cu crossvein as well, but poorly defined.. .... *semifuscata* subgroup (in part)

— Wings usually unmarked (rarely with the dm–cu crossvein faintly brown, but never the anterior margin) .... 4.

4. (3) Front tarsus with elongate cilia only on the first two segments (Fig. 11C). O'ahu .... *D. wikstroemiae* n. sp. (in part)

— Front tarsus with elongate cilia on segments 1–4........ *hirtitarsus* subgroup

5. (2) Body shining black, face usually white; labellum with an appendage that is not much longer than the labellum and fixed, arising from the dorsal end; labellum with strong but inconspicuous black spines; front tarsus and/or tibia often with very long setae (sometimes longer than the basitarsus; Fig. 22)... *setiger* subgroup

— Body yellow to dark brown, pollinose to varying degrees; labellum variable, but not as in Fig. 22A; front leg with only moderately long cilia, never much longer than the basitarsus... 6.

6. (5) Labellum with a sclerotized appendage, usually bearing black setae, or consisting entirely of sclerotized lobes and fleshy portion completely lacking.. 7.

— Labellum without appendages (enlarged dorsoapical setae in some *dissita* subgroup species are superficially similar)... 8.

7. (6) Appendage articulated separately from the labellar sclerite, unbranched; mouthparts otherwise more or less normal, fleshy portion of the labellum present. Kaua'i .... *D. adventitia* Hardy

— Appendage consisting of the enlarged labellar sclerite, articulating only against the labium, often branched; fleshy portion of the labellum absent (Figs. 12A, 13A) .... *ceratostoma* subgroup

8. (6) Labellum with a dense fringe of yellow-white, hair-like setae, longest and sinuate ventrally; dorsally with 5 darker, straighter setae (Fig. 3A). Hawai'i .... 9.

— Hair-like setae of labellum, if present, shorter than enlarged or spine-like setae and straight or slightly curved, never sinuate .... 10.

9. (8) Front tarsus with elongate cilia, more than twice as long as the width of the tarsus (Fig. 3D). Hawai'i (windward side) .... *D. barbata* n. sp.

— Front tarsus with short setulae, hardly longer than the width of the tarsus (Fig. 3E). Hawai'i (Kona and Ka'ū) .... *D. umiumi* n. sp.

10. (8) Labellum with a cluster of 3–5 long, strong, dark brown, inclinate, spine-like setae at the dorsal end, (some usually appressed and appearing as one), strongly differentiated from the remaining setae which are hair-like; at least the 2 dorsal spines closely-placed, separated by less than their width (Figs. 19A, 21A) ....
...*semifuscata* subgroup (in part, including the former *bridwelli* subgroup)

— Labellum with at least some spine-like setae in the medial portion, often grading into hair-like forms; or if similar to the above (e.g., *D. apiki*) then the dorsal spines are relatively widely separated, by more than their width .... 11.

11. (10) Vein R_{2+3} at least faintly infuscated medially, often with more extensive picture-wing like markings .... *fuscoamoeba* subgroup

— Wings hyaline or with marks at the apices of veins and/or over the dm–cu crossvein only .... 12.

12. (11) Mesonotum with 2–4 median pairs of strong acrostichal setae, about 3 times as long as the other acrostichal setulae; front tarsi laterally compressed, rather densely covered with dorsal cilia but nearly bare anteriorly, lacking even straight setulae (Fig. 1C); labellum as in Fig. 1A. Maui .... *D. acrostichalis* Hardy

— Mesonotum lacking strong acrostichals, those in the median rows never more than twice as long as the others (rarely with a single pair elongated); front tarsi variable, but not as described above .... 13.

13. (12) Labellum with both dorsally and ventrally directed elongate, prostrate, round, brown spines (Fig. 7A). Lāna'i .... *D. lelolua* n. sp.
— Labellar spines more or less erect, or if prostrate, they are flattened and all directed dorsally (as in Fig. 15A) .... 14.

14. (13) Labellum with black spines, prostrate or erect; or if yellow to brown (*D. comatifemora*), they are prostrate, closely appressed, and not strongly curved .... 15.
— Labellum with only yellow to brown spines or setae, mostly erect (at least some attached nearly at right angles to the labellar margin) .... 17.

15. (14) Front basitarsus with elongate posterior cilia, but only a few short, inconspicuous dorsal cilia; labellum with two strong, black, dorsal spines, the first (dorsal) strongly curved and over three times as long as the second; aedeagal apodeme very small (only about as long as the basal width of the aedeagus), basal angle obtuse (Fig. 9). Hawai'i .... *D. toxacantha* n. sp.
— Front basitarsus without elongate posterior cilia, or if present then longer dorsal cilia are present as well; dorsal-most labellar spine no more than twice as long as the second .... 16.

16. (15) Labellum with spine-like or scale-like setae, either blunt-tipped and erect or sharply-pointed and appressed, usually 6 or more, dorsoventrally flattened (only 3 and laterally flattened in *D. kualii*), relatively evenly spaced, not clustered dorsally, usually black but sometimes orange-brown (Figs. 15A–17A) .... *freycinetiae* subgroup
— Labellum with 3–4 erect, sharply-pointed, black, spine-like setae clustered dorsally (*D. mitchelli* with 2 additional ventral spines)...............................
.... *mitchelli* subgroup

17. (14) Labellar spines set on prominent lobes of the labellar sclerite ....
.... *scolostoma* subgroup
— Labellar sclerite not lobed .... 18.

18. (17) Labellum with 6 strong, round, brown spines, nearly straight in lateral view, similar in form except one or two shorter than the others (Fig. 14A). Hawai'i (*freycinetiae* subgroup) .... *D. anapuu* n. sp.
— Labellum not as above; if 6 or more spines are present, some are strongly curved or flattened, or they are of two or more distinctly different forms ...
.... 19.

19. (18) At least three ventral labellar setae laterally flattened, blade-like, and conspicuously directed ventrally beyond the ventral margin of the labellum .... *quadrisetae* subgroup

— All labellar setae curved dorsally or medially, or nearly straight (*D. mimica* complex with one seta ventrally-directed)...20.

20. (19) Labellum with 8 or more lateral spine-like setae, usually distinct from hair-like setae by being flattened or bent; if ventral spines are thin, then at least one dorsal spine is strongly broadened and tusk-like .... *dissita* and *mimica* subgroups

— Labellum with 1–6 lateral spines and additional yellow hair-like setae, the two forms grading together; spine-like setae always round, never flattened (except in *D. gladius*) ....21.

21. (20) Tarsal cilia relatively short, less than twice as long as width of basitarsus (except in *D. albifacies*, which has the face white); wing unmarked; epandrium strongly narrowed dorsally, abruptly broadened into a semicircular anterior lobe laterally (which sometimes bears a triangular process anterodorsally); labellum with a strong spine, nearly straight, other spines ventrally grading into hair-like setae (Fig. 18) ......*nanella* subgroup

— Tarsal cilia usually longer, over twice as long as width of basitarsus, or if shorter then wing with the dm–cu crossvein infuscated; epandrium variable, but never very narrow dorsally with a strong anterolateral lobe; labellum with at least two strong, curved spines ....22.

22. (21) Form of labellar sclerite distinctive: almost rectangular, apical margin nearly straight, with a process arising from the ventroapical corner and curving around the ventral portion; labellum with 3 widely-spaced lateral spine-like setae; tarsal cilia short, less than twice as long as the basitarsal diameter (Fig. 8); 5X ratio (apical segment of CuA_1/dm–cu crossvein) 0.8–1.1 ....23.

— Labellar sclerite not as above, narrowing ventrally, apical margin slightly sinuate; labellar setae variable, but if widely-spaced then they are extremely strong spines; tarsal cilia variable in length, sometimes longer; 5X ratio at least 1.2, usually over 1.5 ....24.

23. (22) Aedeagal apodeme convex ventrally (Fig. 8B). O‘ahu ....*D. omnivora* n. sp.

— Aedeagal apodeme concave ventrally (Fig. 8C). Kaua‘i............................ .... *D. wahihuna* n. sp.

24. (22) Labellum with three strong, laterally flattened, widely-separated spines medially and a fan-shaped cluster of 6–7 straight spines ventrally, lacking elongate hair-like setae (Fig. 4A). Hawai'i........................ *D. gladius* n. sp.
— Labellar spines round, grading into hair-like setae..................................25.

25. (24) Tarsus with inconspicuous, relatively short cilia, hardly longer than the prostrate setulae and less than twice as long as the width of the basitarsus. ....26.
— Tarsus with conspicuous, elongate cilia along at least the first two segments, about three times as long as the width of the basitarsus.........27.

26. (25) Palpi yellow, with several elongate setae along the lateral margin but no strong apical seta; epandrium parallel-sided; aedeagal apodeme not quite as long as high (Fig. 10B). Maui..*D. waikamoi* n. sp.
— Palpi dark brown with a strong apical seta; epandrium distinctly narrower along the dorsal margin than laterally; aedeagal apodeme about 1.5 times as long as high (Fig. 6B). Maui ....*D. komohana* n. sp.

27. (25) Front tarsus with elongate cilia only on the first two segments; labellum with only two thin spines, often not distinguishable from hair-like setae in pinned specimens (Fig. 11A). O'ahu ....*D. wikstroemiae* n. sp. (in part)
— Front tarsus with elongate cilia on at least the first three segments; labellum with at least 3–4 moderately strong spines ....25.

28. (27) Face, palpi, and coxae dark brown; front basitarsus with both anterodorsal and posterodorsal cilia; aedeagus nearly linear, apodeme much longer than high (Fig. 5). Maui.. *D. incongruens* n. sp.
— Face, palpi, and coxae pale yellow; front basitarsus with only anterodorsal cilia; aedeagus arched, basal angle weak, apodeme about as long as high (Fig. 2). Kaua'i .... *D. apiki* n. sp.

SPECIES ACCOUNTS

unplaced species

The following species do not fit clearly into the currently-defined subgroups, but do not appear to be distinctive or diverse enough to warrant the creation of new subgroups. Three previously-described species, *D. acrostichalis*, *D. adventitia*, and *D. tetraspilota*, are also considered unplaced; the second is included in the key due to its mouthpart modifications but is probably more closely related to the *nudidrosophila* and *picture wing* subgroups based on the form of the ovipositor and DNA sequence data (Bonacum 2001). The male of *D. tetraspilota* is unknown, but it is likely to be a *modified mouthparts* species based on the form of the ovipositor and the wing marks.

Of the new species, *D. toxacantha* is superficially similar to species of the *mitchelli* subgroup due to its black labellar spines; however, the tarsal ciliation and genitalia are very different, and mtDNA sequences indicate that it is much more closely related to members of the *mimica* subgroup (unpubl. data). *Drosophila barbata* and *D. umiumi* do not resemble any other species in collections. The presence of several strong, dark, dorsal setae followed by numerous pale, thin ventral setae on the labellum is suggestive of the *bridwelli* subgroup, while the genitalia and host usage patterns are closer to the *dissita* subgroup. Likewise, the species pair of *D. omnivora* and *D. wahihuna* are sister taxa but are not close to any others; the mouthparts and ecological records bear some resemblance to the *fuscoamoeba* subgroup (although the rectangular labellar sclerite appears to be unique), while the genitalia are closer to the *scolostoma* subgroup. The mouthparts of *D. lelolua* are somewhat suggestive of *D. comatifemora* of the *freycinetiae* subgroup, but the setae are not as conspicuously flattened and scale-like as typical for the subgroup, and no other members have prostrate, strongly ventrally-directed setae as *D. lelolua* does. Discovery of the male of *D. acrostichalis* shows it to be very different from the *semifuscata* subgroup, where it was previously included, but not particularly close to any others; likewise, the mouthparts of *D. gladius* show little similarity to other species.

The remaining species – *D. apiki*, *D. incongruens*, *D. komohana*, *D. waikamoi*, and *D. wikstroemiae* – are largely distinguished by lacking the characters diagnostic for the other species and subgroups. The mouthparts generally appear to be intermediate between the *fuscoamoeba*, *nanella*, *scolostoma*, and *semifuscata* subgroups, with a few simple, spinose setae, and often need to be dissected to reveal their true form. *Drosophila incongruens* is the species referred to as "*Drosophila* n. sp." in Hardy and Kaneshiro's (1975b) review of the *mitchelli* subgroup, but based on the mouthparts and genitalia it does not belong there. Until more detailed phylogenetic work is done to elucidate their true relationships, it seems best to leave all these species unplaced.

Drosophila acrostichalis Hardy
Fig. 1

Drosophila acrostichalis Hardy, 1965:132–133. Holotype ♀, Maui, Haleakalā Crater, Palikū, viii.1952, W. C. Mitchell. BPBM 6285 [examined].

DIAGNOSIS. A relatively large, dark-colored species. Both the male and female are readily distinguished by the presence of strong acrostichal setae. The male mouthparts and front tarsus are also unique.

DESCRIPTION. ♂. **Head.** Front velvety brown; fronto-orbital plates and ocellar triangle dark brown, conspicuously pollinose. Anterior reclinate seta slightly posterior of the proclinate, about 2/3 as long; posterior reclinate about 2.5 times as long as the anterior. Ocellar setae slightly shorter than the posterior reclinate. Vertical setae normal in position, about as long as the posterior reclinate. Eyes with short, inconspicuous pile, about as long as one facet. Face dark brown, with a median carina. Antenna dark brown, first segment paler; arista with 5–6 dorsal and 2 ventral rays in addition to the apical fork. Several elongate oral setae present at the vibrissal angle (slightly longer than the posterior oral setulae, which are also relatively long), but no distinctly stronger oral vibrissae. Gena yellow dorsally, this portion strongly narrowed anteriorly due to the expanded brown margin (anatomically part of the face). Palp pale brown, broad, flattened, setulose, broadest near the apex which is obliquely truncate; unusually short, about as wide as long on the inner margin (outer margin longer); no distinct apical setae. Labellar sclerite short and unusually broad, strongly sclerotized; dorsally with several strong setae and a large, flattened, apically-directed spine, the latter partially concealed behind two long dorsally-directed spines in lateral view; laterally with 5 short, broad, flattened, scimitar-shaped spines; ventrally with a bare, emarginate section followed by another long, flattened spine and a transverse row of 6 round spines, the latter visible only in ventral view (Fig. 1A). **Thorax.** Entirely brown and pollinose, except for the humerus which is mottled with yellow; mesonotum with three faint gray stripes, medially and along the dorsocentral rows. Two humeral setae, the ventral nearly as long as the dorsal. Acrostichal setulae in about 8–10 irregular rows; 2–4 pairs in the median rows at the level of the notopleuron are enlarged, about as long as the proclinate fronto-orbital seta and 3 times as long as the normal acrostichals, gradually becoming smaller posteriorly. Two pairs of dorsocentral setae, the anterior about half as long as the posterior and set rather far back, distinctly posterior of the second supraalar seta. Halteres pale brown, stems darker. **Legs.** Pale yellowish brown, femora and hind tarsus darker. Front tibia lacking ciliation. Front basitarsus slightly swollen, short, only about 1/3 as long as the tibia; dorsally densely covered with long, strong cilia (Fig. 1C). Front tarsal segments 2–4 distinctly laterally compressed, with a row of strong, more or less sinuate cilia along the dorsal ridge; tarsus bare on anterior surface, lacking setulae. **Wings.** Infuscated anteroapically along veins R_{2+3} and R_{4+5} up to about the level

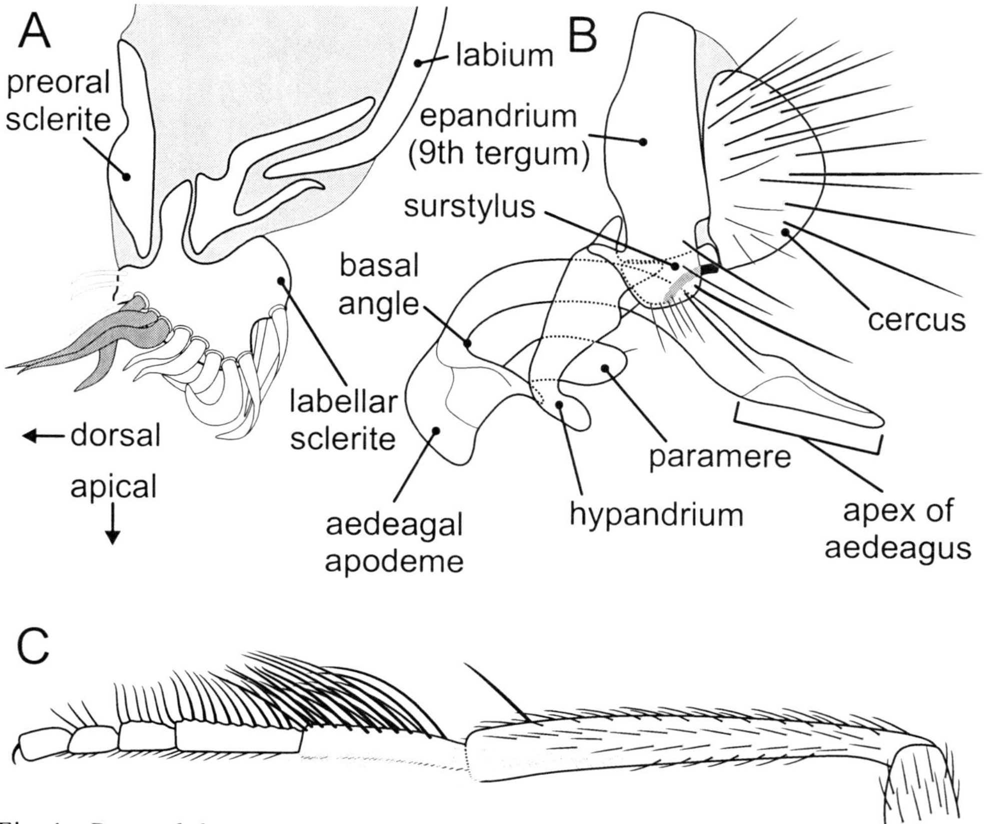

Fig. 1. *Drosophila acrostichalis* male. (A) Lateral view of labellum. (B) Terminalia. (C) Right front leg, anterior view.

of the dm–cu crossvein, about half as far along vein M; a large spot present over the dm–cu crossvein as well (see Fig. 22a in Hardy 1965). Costal fringe extending about half the distance between the apex of R_{2+3} and R_{4+5}. **Abdomen.** Entirely dark brown. Epandrium slightly narrowed dorsally; anteroventral spine curled under, thus appearing narrow in lateral view. Aedeagal apodeme about as long as high, nearly straight ventrally. Aedeagus narrow at the base, broader beyond, strongly bent, basal angle acute, apex elongate; preapical protuberance small, low (Fig. 1B). ♀. Identical to the male with the following exceptions. **Head.** Palp brown, about twice as long as wide, apex broadly rounded, with a strong subapical seta. One to two elongate oral vibrissae present. Labellum unmodified. **Legs.** Brown. Front legs without elongate cilia or setae. **Abdomen.** Ovipositor yellow, narrow, parallel-sided.

MEASUREMENTS. N = 4♂. TL = 2.06 (2.02–2.11) mm; WL = 4.34 (4.21–4.42) mm; TL/WL = 0.5; HW = 1.31 (1.25–1.33) mm; HW/FS = 2.0; HW/TL = 0.6 (0.6–0.7); CI = 5.1 (4.9–5.3); 4V = 1.3 (1.2–1.3); 5X = 1.3 (1.2–1.4); 4C = 0.4 (0.4–0.5); M = 0.4 (0.3–0.4). N = 2♀. TL = 1.96 (1.95–1.97) mm; WL = 4.17 (4.16–4.17) mm;

TL/WL = 0.5; HW = 1.20 (1.19–1.22) mm; HW/FS = 2.0 (1.9–2.0); HW/TL = 0.6; CI = 4.7 (4.6–4.9); 4V = 1.2 (1.2–1.3); 5X = 1.3 (1.3–1.4); 4C = 0.4; M = 0.4.

MATERIAL EXAMINED. MAUI: 2♂, ʻĀinahou Valley, 29–30.iii.1970, P16, KYK. 2♂ 3♀, Palikū, 9.x.1975, KYK. 1♂, Kaupō Gap, 9.x.1975, KYK (all UHIM).

DISTRIBUTION & ECOLOGY. Known from East Maui in the vicinity of Haleakalā Crater.

DISCUSSION. This species was tentatively placed in the *semifuscata* subgroup (Magnacca and O'Grady 2007) based on the similarity of the wing markings and body coloration. However, discovery of the male has revealed that the mouthparts are very different. It does not clearly fit with any other species, but appears to show some similarity to the *dissita/mimica* complex, based on the presence of laterally flattened spines and their division into dorsal, lateral, and ventral clusters. The ovipositor is more like that of the *nudidrosophila* and *picture wing* groups than other *modified mouthparts* species, but is held horizontally rather than vertically (see Fig. 22b in Hardy, 1965).

Drosophila adventitia Hardy

Drosophila adventitia Hardy, 1965:136–138. Holotype ♂, Poʻomau Valley, 3000 ft., vii.1952, DEH, BPBM 6287 [examined].

DIAGNOSIS. Easily separated from all other Hawaiian *Drosophila* by the unique labellum of the male, which possesses a moveable appendage at the dorsal end (in contrast to members of the *ceratostoma* and *setiger* subgroups, where the appendage is an extension of the labellar sclerite and fixed to it) while retaining the fleshy portion.

DESCRIPTION. ♂. **Head.** Front brown, paler anteriorly, pollinose; fronto-orbital plates and ocellar triangle slightly darker. Anterior reclinate seta slightly posterior of the proclinate, about half as long; posterior reclinate about 3 times as long as the anterior. Ocellar setae slightly longer than the posterior reclinate. Vertical setae normal in position, about equal to the posterior reclinate. Eyes with short, inconspicuous pile, about as long as one facet. Face brown, paler medially; almost flat, only weakly carinate medially. Antenna brown; arista with 6–7 dorsal and 2–3 ventral rays in addition to the apical fork. No strong oral vibrissa. Gena largely yellow, brown near the eye. Palp yellowish brown, flattened, clavate, setulose; with 4–6 longer setulae near the apex but no stronger apical or subapical setae. Labellum bearing an elongate, articulated, rod-like appendage at the dorsal end; otherwise normal, without spinose setae or other modifications. Appendage about as long as antenna, thin, and capitate, with several short setae at the apex, two strong, black, hair-like and two pale, flattened, and scale-like; also with several short setulae along its length (see Fig. 1e in Magnacca and O'Grady 2007). **Thorax.** Dark

brown, lightly pollinose but shining; yellow near the humeri. Two humeral setae, the ventral about 1/2–3/5 as long as the dorsal. Acrostichal setulae in 6–8 irregular rows. Only one strong dorsocentral seta; the anterior is distinguishable on close inspection but is very weak, less than 1/3 as long as the posterior and only slightly longer than the other setulae. Halteres yellow. **Legs.** Femora dark brown, coxae slightly paler; tibiae largely yellow, banded with brown at the bases and apices, sometimes inconspicuously so; tarsi yellow. Front tibia lacking ciliation. Front tarsus with an anterodorsal row of short cilia along the entire length; these mostly only about as long as the width of the tarsus, slightly longer apically. Front basitarsus also with 3 longer, widely-spaced anteroventral cilia along the apical half. **Wings.** Hyaline, faintly infuscate but without distinct markings. Costal fringe extending about halfway between the apex of R_{2+3} and R_{4+5}. **Abdomen.** Entirely brown.
♀. Identical to the male with the following exceptions. **Head.** Palp widest near the middle, with a strong apical seta. Labellum unmodified. **Thorax.** Two strong dorsocentral bristles present, the anterior about 3/4 as long as the posterior. **Legs.** Front legs without cilia. **Abdomen.** Ovipositor yellow, elongate and narrow, with a constricted apex.

MATERIAL EXAMINED. KAUA'I: 1♂, same data as holotype. 2♂, No. Fork Wailua River, 1200 ft., 12.iii.1968, KYK. 1♂ 3♀, Honopū, ex *Charpentiera* stem, xi.1978, SLM. 1♂, Kawaikai [Kawaikōī?], ex *Charpentiera* stem, SLM (all UHIM).

DISTRIBUTION & ECOLOGY. Kaua'i. Reared from stems of *Charpentiera* (pāpala, Amaranthaceae).

DISCUSSION. The male mouthparts and front tarsus were illustrated by Hardy (1965, Fig. 24). The female is described here for the first time, from specimens reared in association with a male. They are significant because the form of the ovipositor is similar to that of the *picture wing* and *nudidrosophila* groups, rather than other *modified mouthparts* species. This information is congruent with the phylogenetic results of Bonacum (2001), who found that *D. adventitia* + *D. primaeva* form the sister group to the *picture wing/nudidrosophila* clade. In addition, the mouthparts of *D. adventitia* are also unlike those of any other known *modified mouthparts* group species. While Bonacum's (2001) results are preliminary and need to be confirmed by more intensive analyses, the sum of data suggest that *D. adventitia* does not phylogenetically belong in the *modified mouthparts* group.

Drosophila apiki **n. sp.**
Fig. 2

DIAGNOSIS. A relatively nondescript species, distinguished by the combination of elongate front tarsal cilia, four moderately strong dorsal labellar spines, and an arched aedeagus.

DESCRIPTION. ♂. **Head.** Front brown, pale anteriorly; fronto-orbital plates and ocellar triangle darker and conspicuously pollinose. Anterior reclinate seta slightly posterior of the proclinate, about 2/3 as long; posterior reclinate about 3 times as long as the anterior. Ocellar setae about as long as the posterior reclinate. Vertical setae normal in position, slightly longer than the posterior reclinate. Eyes with short, inconspicuous pile, about as long as one facet. Face pale yellow, with a rounded carina medially. Antenna pale brown; arista with 5–6 dorsal and 2 ventral rays in addition to the apical fork. One strong oral vibrissa, the next seta 1/2–2/3 as long, remaining oral setulae smaller. Gena yellow. Palp light brown, flattened, broadest near the middle, setulose; with an elongate apical seta and several additional elongate setae along the lateral margin. Labellum notched near the middle; dorsally with 4 yellow spine-like setae, strongest at the dorsal end; ventrally with a total of about 10 more hair-like setae in two rows (Fig. 2A). **Thorax.** Brown, pollinose; katepisternum pale. Two humeral setae, the ventral nearly as long as the dorsal. Acrostichal setulae in 6–8 irregular rows. Two pairs of dorsocentral setae, the anterior about 2/3 as long as the posterior. Halteres yellow-brown. **Legs.** Light brown. Front tibia lacking ciliation. Front basitarsus with about 8 elongate anterodorsal cilia in two irregular rows along the entire length; second tarsal segment with 3–4 cilia and remaining segments with 1–2 each (Fig. 2C). **Wings.** Hyaline, faintly infuscated along R_{2+3}. Costal fringe extending 1/3 the distance between the apex of R_{2+3} and R_{4+5}. **Abdomen.** Entirely brown. Epandrium apparently broader along the dorsal margin than at the ventral lobe, anteroventrally with a broad spine. Aedeagal apodeme moderately large, semiovate. Aedeagus weakly bent at the base, evenly curved beyond, becoming slightly wider between the apodeme and apical membranous region, apex narrow and elongate; preapical protuberance absent (Fig. 2B).

♀. Identical to the male with the following exceptions. **Head.** Labellum unmodified. **Legs.** Front legs without elongate cilia or setae. **Abdomen.** Ovipositor brown, broad and blunt.

MEASUREMENTS. N = 1♂. TL = 1.32 mm; WL = 2.82 mm; TL/WL = 0.5; HW = 1.02 mm; HW/FS = 2.3; HW/TL = 0.8; CI = 3.8; 4V = 1.3; 5X = 1.6; 4C = 0.6; M = 0.4. N = 1♀. TL = 1.47 mm; WL = 3.15 mm; TL/WL = 0.5; HW = 1.08 mm; HW/FS = 2.3; HW/TL = 0.7; CI = 3.8; 4V = 1.4; 5X = 1.6; 4C = 0.6; M = 0.4.

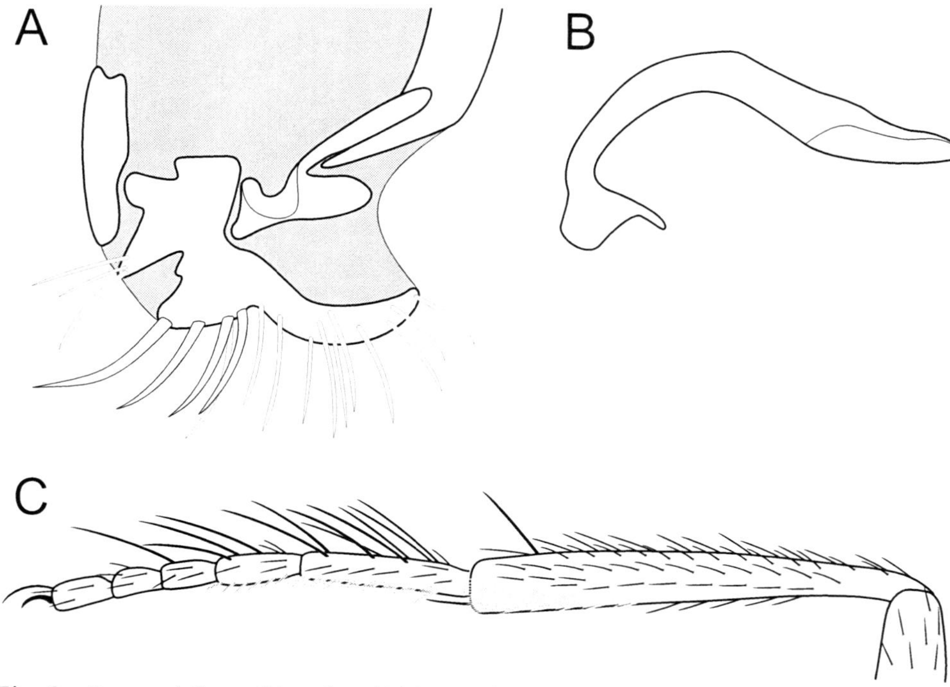

Fig. 2. *Drosophila apiki* male. (A) Lateral view of labellum. (B) Aedeagus. (C) Right front leg, anterior view.

TYPES. KAUA'I: Holotype ♂ (BPBM 16710) and allotype ♀, Kahili, *Labordia* stem, SLM [no other information].

DISTRIBUTION & ECOLOGY. Kaua'i; known only from the type collection. Reared from stems of *Labordia* sp. (kamakahala, Loganiaceae).

ETYMOLOGY. From the Hawaiian *'āpiki*, peculiar, strange, or deceitful, in reference to both its use of *Labordia* as a host (the only such record for Hawaiian Drosophilidae), and the difficulty in discerning its relationships.

DISCUSSION. The male terminalia of this species, particularly the epandrium, are difficult to interpret as they were apparently inverted due to drying of the specimen; as a result, only the aedeagus is illustrated.

Drosophila barbata **n. sp.**
Fig. 3A–D

DIAGNOSIS. Readily separated from other species by the cluster of tangled white to yellow setae on the labellum, with 5 darker setae dorsally. Very similar to *D. umiumi*, distinguished by having elongate cilia on the front tarsus.

DESCRIPTION. ♂. **Head.** Front brown, pale anteriorly; fronto-orbital plates and ocellar triangle darker and conspicuously pollinose. Anterior reclinate seta about even with the proclinate, about 2/3 as long; posterior reclinate about 3 times as long as the anterior. Ocellar setae about as long as the posterior reclinate. Vertical setae normal in position, about as long as the posterior reclinate. Eyes with short, inconspicuous pile, about as long as one facet. Face pale yellowish brown, with a rounded carina medially. Antenna brown; arista with 5–6 dorsal and 3 ventral rays in addition to the apical fork. Several elongate oral setae present, but no distinct oral vibrissae. Gena yellow, narrowly brown along the oral margin. Palp dark brown, flattened, weakly clavate, broadest near the apex, setulose; with 2 strong setae, one apical and one subapical (occasionally a second subapical seta present). Labellum with three distinct areas of sclerotization; the first bearing 5–6 curved brown setae laterally, and the second with 3 elongate, somewhat sinuate, white setae; all with numerous white inclinate setae medially (Fig. 3A, B). Preoral sclerites unusually dark and prominent. **Thorax.** Brown, pollinose, with a yellow stripe along the anapleural suture; the humerus and center of the anepisternum also tinged with yellow. Two humeral setae, the ventral about 3/4 as long as the dorsal. Acrostichal setulae in 6 rows. Two pairs of dorsocentral setae, the anterior about 3/4 as long as the posterior. Halteres yellow, stems brownish. **Legs.** Yellow, tinged faintly with brown. Front tibia lacking ciliation. Front basitarsus with 8–10 anterodorsal cilia in two irregular rows along the entire length; second tarsal segment with 4–5 cilia and remaining segments with 1–2 each (Fig. 3D). **Wings.** Entirely hyaline (sometimes faintly infuscated at the apices of the veins). Costal fringe extending 2/5–1/2 the distance between the apex of R_{2+3} and R_{4+5}. **Abdomen.** Brown, paler on the basal half of each segment. Epandrium slightly broader along the dorsal margin than at the ventral lobe, anteroventrally with a broad spine; dorsal margin distinctly peaked about 1/3 of the way along. Aedeagal apodeme small, rounded triangular. Aedeagus narrow at the base, evenly curved and fishhook-shaped, apex broad and blunt; preapical protuberance absent (Fig. 3C).
♀. Identical to the male with the following exceptions. **Head.** One or two distinct oral vibrissae present. Labellum unmodified. **Legs.** Front legs without elongate cilia or setae. **Abdomen.** Ovipositor yellow, broadly triangular, apex blunt.
MEASUREMENTS. N = 5♂. TL = 1.20 (1.09–1.28) mm; WL = 2.54 (2.40–2.64) mm; TL/WL = 0.5; HW = 0.93 (0.86–1.00) mm; HW/FS = 2.1 (2.0–2.2); HW/TL = 0.8 (0.7–0.8); CI = 4.2 (4.1–4.4); 4V = 1.6 (1.5–1.7); 5X = 1.7 (1.6–2.0); 4C = 0.6 (0.5–0.6); M = 0.5 (0.4–0.5). N = 1♀. TL = 1.42 mm; WL = 2.76 mm; TL/WL =

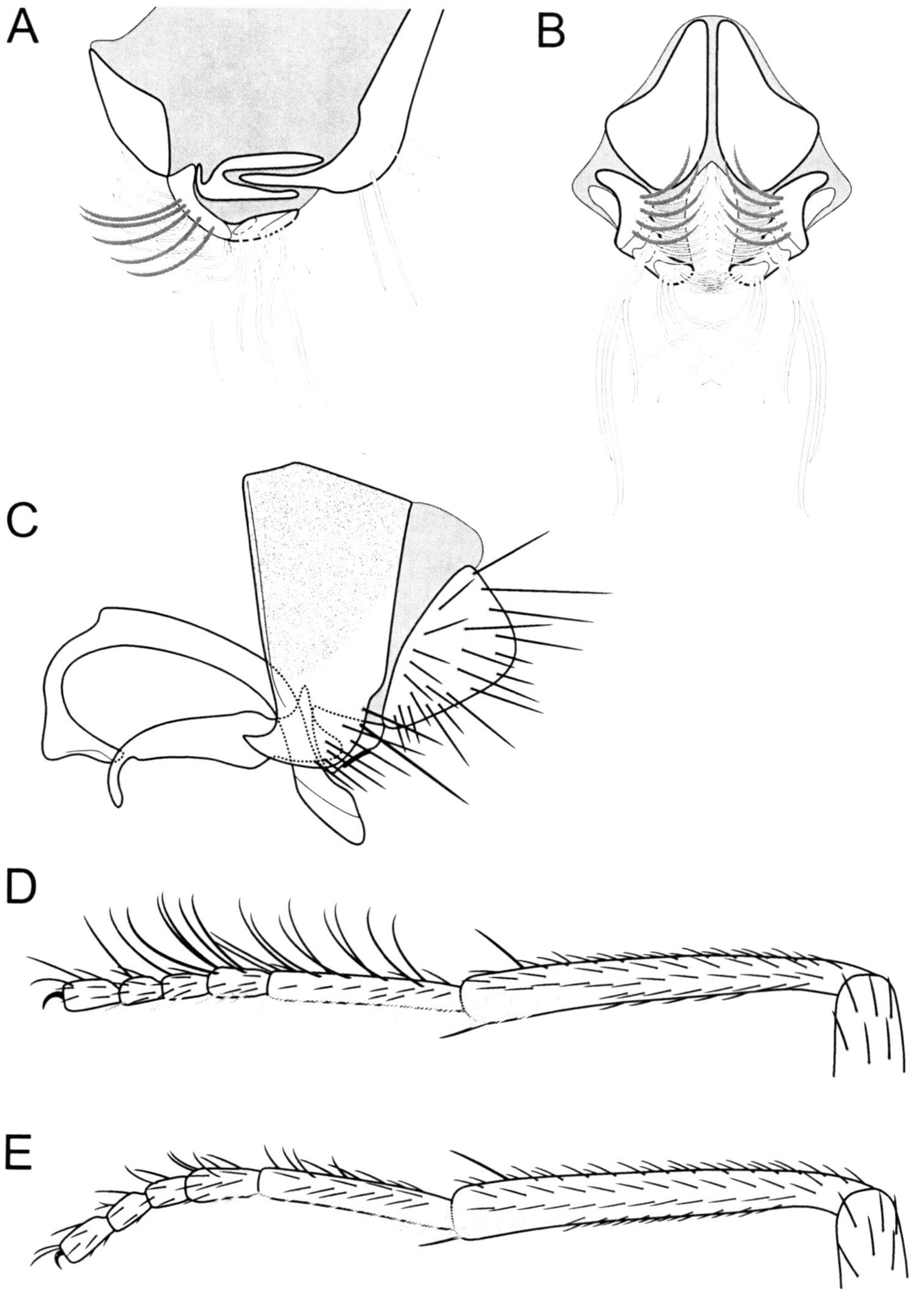

Fig. 3. *Drosophila barbata* male. (A) Lateral and (B) dorsal view of labellum. (C) Terminalia. (D) Right front leg, anterior view. (E) *Drosophila umiumi* male, right front leg, anterior view.

0.5; HW = 1.02 mm; HW/FS = 2.2; HW/TL = 0.7; CI = 4.5; 4V = 1.7; 5X = 1.6; 4C = 0.6; M = 0.5.

TYPES. HAWAIʻI: Holotype ♂ (BPBM 16716) and allotype ♀, Saddle Rd., 2275 ft., reared ex *Clermontia* stems, G92A, 6.vii.1966, WBH.

PARATYPES. HAWAIʻI: 1♂, forest no. of Puʻu Kapu, 2800 ft., 24.viii.1963, LHT. 7♂, 9.4 mi. above Hilo, 2275 ft., reared ex *Clermontia* stems, G69A, 2.vi.1966, WBH. 2♂, Kohala Mts., ʻĀwini Camp, 1–2.viii.1966, KYK. 1♂, ʻĀwini Cabin on Kohala Ditch Trail, reared ex rotting *Clermontia* stems, 1.viii.1966, HLC. 1♂, ʻŌlaʻa, ex *Charpentiera* stem, viii.1974, SLM (all UHIM).

DISTRIBUTION & ECOLOGY. East Hawaiʻi: Puna, Hilo, and North Kohala districts; probably occurs in Hamākua as well. Widespread in wet forest. Reared primarily from bark and stems of *Clermontia* spp. (ʻohe wai, Campanulaceae) and *Urera glabra* (ōpuhe, Urticaceae); also single records from *Charpentiera obovata* (pāpala, Amaranthaceae), *Pisonia brunoniana* (pāpala kēpau, Nyctaginaceae) and *Freycinetia arborea* (ʻieʻie, Pandanaceae) (Magnacca et al. 2008).

ETYMOLOGY. From the Latin *barbatus*, bearded, in reference to the appearance of the labellar setae.

DISCUSSION. Although this species is relatively commonly reared from a variety of plants, there are very few specimens from hand collecting or baits. *Drosophila barbata* and *D. umiumi* are both known from relatively large series from multiple localities, and show a distinct geographic difference – *D. barbata* with long tarsal cilia in Kohala, Hilo, and Puna, and *D. umiumi* with short tarsal cilia in Kaʻū and Kona – with no intermediates. Therefore, we feel confident in describing them as separate species. The parameres are not clearly visible in any dissected specimens of either species.

Drosophila gladius **n. sp.**

Fig. 4

DIAGNOSIS. The scimitar-shaped dorsal spine of the labellum is diagnostic for this species; it somewhat resembles the tusk-like spines of the *mimica* or *dissita* subgroups, but is nearly straight, dorsally-directed, and there is a total of only three strong lateral spines.

DESCRIPTION. ♂. **Head.** Front brown, pale anteriorly; fronto-orbital plates and ocellar triangle darker and conspicuously pollinose. Anterior reclinate seta slightly posterior of the proclinate, about 1/3 as long; posterior reclinate about 4 times as long as the anterior. Ocellar setae about as long as the posterior reclinate. Vertical setae normal in position, about as long as the posterior reclinate. Eyes with short, inconspicuous pile, about as long as one facet. Face brown, with a rounded carina medially. Antenna brown; arista with 4–5 dorsal and 2–3 ventral rays in addition to

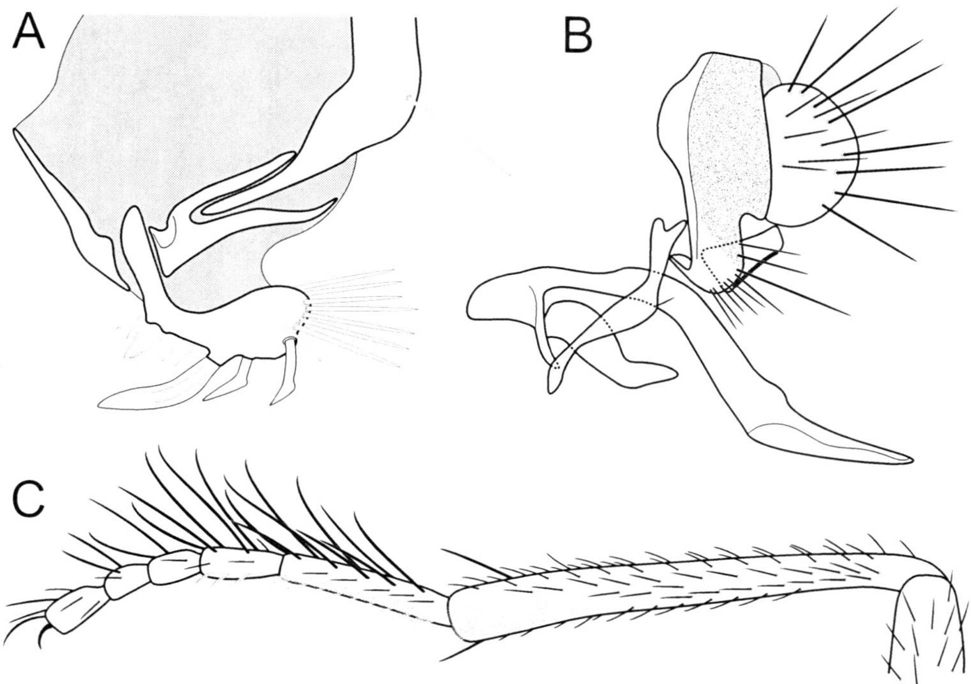

Fig. 4. *Drosophila gladius* male (holotype from Kahuku). (A) Lateral view of labellum. (B) Terminalia. (C) Right front leg, anterior view.

the apical fork. No distinct oral vibrissae or elongate setae. Gena yellow, narrowly brown along the oral margin. Palp yellow, flattened, broadest in the apical 1/4, sparsely setulose; with an elongate subapical seta and several additional elongate setae along the lateral margin. Labellum with 3 large, brown, laterally flattened, spine-like setae, the first sword-shaped, broad, and elongate, the other two shorter and narrower; ventrally with about 7 more hair-like setae in a fan arrangement (Fig. 4A). **Thorax.** Mesonotum and most of upper half of pleura brown, lightly pollinose; katepisternum yellow. Two humeral setae, the ventral about 3/4 as long as the dorsal. Acrostichal setulae in 6–8 rows. Two pairs of dorsocentral setae, the anterior about 1/2–2/3 as long as the posterior. Halteres yellow-brown. **Legs.** Yellow. Front tibia lacking ciliation. Front basitarsus with 9 anterodorsal cilia in two irregular rows along the entire length; second tarsal segment with 5 cilia and remaining segments with 1–2 each (Fig. 4C, holotype; Kohala specimen with 6 cilia along the apical 2/3 of the basitarsus and 3 on the second segment). **Wings.** Entirely hyaline. Costal fringe extending 1/3 the distance between the apex of R_{2+3} and R_{4+5}. **Abdomen.** Entirely brown. Epandrium almost parallel-sided, slightly broadened medially, almost entirely microtrichose; anteroventral spine curled under, thus appearing narrow in lateral view. Aedeagal apodeme longer than high, concave ventrally. Aedeagus narrow at the base, moderately bent, basal angle almost perpendicular, apex elongate; preapical protuberance very small, indistinct (Fig. 4C).

♀. Unknown.
MEASUREMENTS. N = 2♂. TL = 1.13 (0.95–1.31) mm; WL = 2.73 (2.24–3.23) mm; TL/WL = 0.4; HW = 1.00 (0.87–1.14) mm; HW/FS = 2.1; HW/TL = 0.9; CI = 4.2 (3.9–4.5); 4V = 1.6; 5X = 1.7 (1.6–1.8); 4C = 0.6; M = 0.5.

TYPE. HAWAI'I: Holotype ♂ (BPBM 16711), Kahuku Ranch, 4000 ft., 14.iv.1976, DEH.

PARATYPE. HAWAI'I: 1♂, Kohala Mts., Pu'u La'ala'au, M69, 21.viii.1969, KYK (UHIM).

DISTRIBUTION & ECOLOGY. Hawai'i; rare but apparently widespread. Breeding habits unknown.

ETYMOLOGY. From the Latin *gladius*, sword, referring to the shape of the dorsal labellar spine. It is to be considered a noun in apposition.

DISCUSSION. The two specimens included in the type series exhibit slight differences in the genitalia, legs, and mouthparts, and may represent windward/leeward sibling species similar to *D. barbata* and *D. umiumi*. Compared to the holotype from Kahuku, the Kohala specimen has a more elongate and concave aedeagal apodeme; the parameres are straight rather than bent dorsally near the apex; the ciliation of the front leg is slightly different; and the labellum lacks a spinose process medial of the third lateral spine. However, because the differences are relatively subtle, both are from the same island, and only one specimen of each morphotype is available, they are being included as a single species.

Drosophila incongruens **n. sp.**
Fig. 5

DIAGNOSIS. Similar to species of the *mitchelli* subgroup in having several moderately strong, curved, dark labellar spines; distinguished by having the spines brown, evenly spaced, and grading into hair-like setae rather than black and in distinct clusters. Also, the entire body is dark brown, including the femora.

DESCRIPTION. ♂. **Head.** Front brown, paler anteriorly; fronto-orbital plates and ocellar triangle darker and conspicuously pollinose. Anterior reclinate seta slightly posterior of the proclinate, 2/3–3/4 as long; posterior reclinates broken off in both specimens. Ocellar setae about 1.5 times as long as the proclinate. Vertical setae normal in position, over 1.5 times as long as the proclinate. Eyes with short, inconspicuous pile, about as long as one facet. Face dark brown, median carina very low. Antenna brown, 3rd segment darker; arista with 6 dorsal and 2 ventral rays; 2 additional long rays also present near the middle, slightly medial of the dorsal rays in addition to the apical fork. No distinct oral vibrissae, all oral setulae moderately long. Gena narrowly brown along the oral margin, a broader yellow stripe above, brown posterodorsally. Palp light brown, flattened, broadest at the

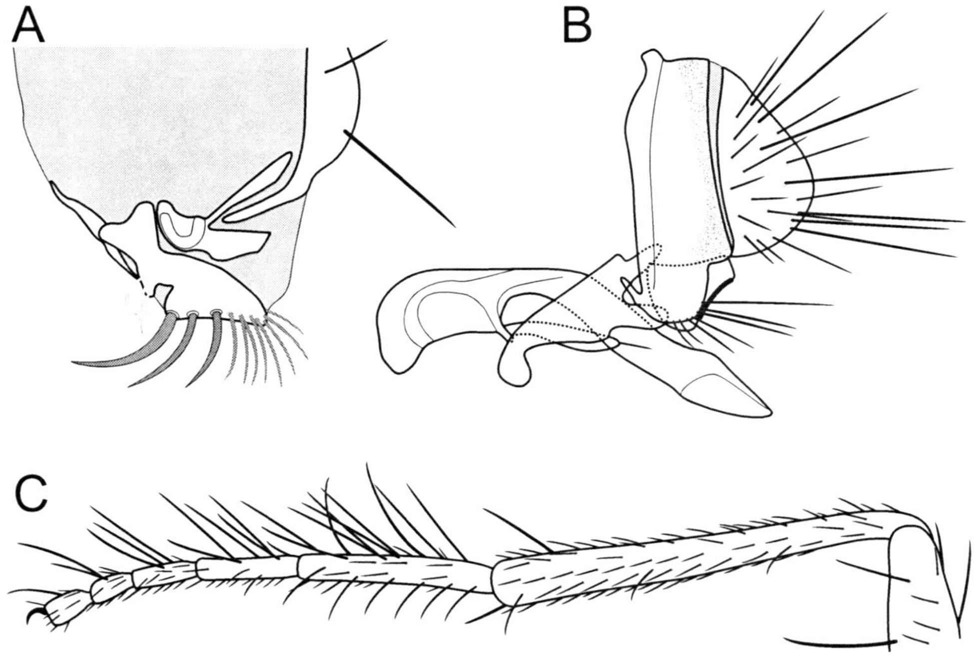

Fig. 5. *Drosophila incongruens* male. (A) Lateral view of labellum. (B) Terminalia. (C) Left front leg, posterior view.

apex, setulose; with an elongate apical seta. Labellum with 4 brown curved, round, spine-like setae, strongest at the dorsal end and grading into the hair-like ventral setae, of which there are about 7 (Fig 5A). **Thorax.** Entirely dark brown, pollinose. Two humeral setae, about equal in length; an enlarged dorsal setula also present, usually about twice as long as other setulae but only about 1/3 as long as the main setae. Acrostichal setulae in 6–8 irregular rows. Two pairs of dorsocentral setae, the anterior about 3/4 as long as the posterior. Halteres brown. **Legs.** Coxae and femora dark brown, tibiae and tarsi brownish yellow. Front tibia with about 3–5 erect, curled, posterior cilia; rather short, only slightly longer than the straight setulae, and inconspicuous. Front basitarsus with about 8 elongate anterodorsal cilia along the entire length and 3–4 dorsal cilia over the apical half, and 6 shorter posteroventral cilia; second tarsal segment with 5–6 anterodorsal to dorsal cilia, third with 3, and remaining segments with 1–2 each (Fig. 5B). **Wings.** Entirely hyaline. Costal fringe extending about halfway between the apex of R_{2+3} and R_{4+5}. **Abdomen.** Entirely dark brown. Epandrium almost parallel-sided, slightly narrowed dorsally; anteroventral spine curled under, thus appearing narrow in lateral view. Aedeagal apodeme longer than high, concave ventrally. Aedeagus narrow at the base, strongly bent, basal angle acute, apex short and moderately broad; preapical protuberance very small, indistinct (Fig. 5C).

♀. Unknown.

MEASUREMENTS. N = 2♂. TL = 1.57 (1.56–1.59) mm; WL = 3.49 mm; TL/WL = 0.5 (0.4–0.5); HW = 1.14 mm; HW/FS = 2.1; HW/TL = 0.7; CI = 3.9 (3.9–4.0); 4V = 1.2 (1.2–1.3); 5X = 1.3 (1.2–1.5); 4C = 0.5; M = 0.3 (0.3–0.4).

TYPE. MAUI: Holotype ♂ (BPBM 16712), Waikamoi, 8.vii.1964, WBH.

PARATYPE. MAUI: 1♂, Waikamoi, 1.x.1964, S21, HTS (UHIM).

DISTRIBUTION & ECOLOGY. Maui; known only from Waikamoi. Breeding habits unknown.

ETYMOLOGY. From the Latin *incongruens*, asymmetrical, referring to the humeral setae.

DISCUSSION. This is the species referred to as "n. sp." by Hardy and Kaneshiro (1975b). Although the three humeral setae of the 8.vii.1964 specimen were used as a defining character in that paper, the right humerus clearly has only two strong setae. The paratype, which apparently was not recognized as conspecific at the time, has two setae on both sides.

Drosophila komohana n. sp.

Fig. 6

DIAGNOSIS. Similar to *D. waikamoi* in having relatively short front tarsal cilia, only about twice as long as the diameter of the basitarsus; separated from that species by having the palpi dark brown rather than yellow and details of the mouthparts and genitalia.

DESCRIPTION. ♂. **Head.** Front dark brown, slightly paler and more pollinose on fronto-orbital plates and ocellar triangle. Anterior reclinate seta slightly posterior of the proclinate, about half as long; posterior reclinate about 3 times as long as the anterior. Ocellar setae about as long as the posterior reclinate. Vertical setae normal in position, about as long as the posterior reclinate. Eyes with short, inconspicuous pile, about as long as one facet. Face dark brown, with a rounded carina. Antenna dark brown, paler on the dorsolateral surface of the second segment; arista with 5 dorsal and 2 ventral rays in addition to the apical fork. Several elongate oral setae present at the vibrissal angle (about twice as long as the posterior oral setulae), but no distinct oral vibrissae. Gena yellow, narrowly dark brown along the oral margin. Palp dark brown, flattened, broadest near the middle, setulose; with an elongate apical seta. Labellum with 3 brown, curved, round, spine-like setae, strongest at the dorsal end and relatively closely-placed; followed by 2 more-widely-spaced spines, and several hair-like setae at the ventral apex of the labellar sclerite (Fig. 6A). **Legs.** Yellow. Front tibia lacking ciliation. Front basitarsus with about 7 dorsal cilia in two irregular rows over the apical 2/3; second tarsal segment with 4–5 dorsal cilia, and remaining segments with 1–2 each; all relatively short, about twice as long as the width of the basitarsus (Fig. 6C). **Wings.** Mostly hyaline,

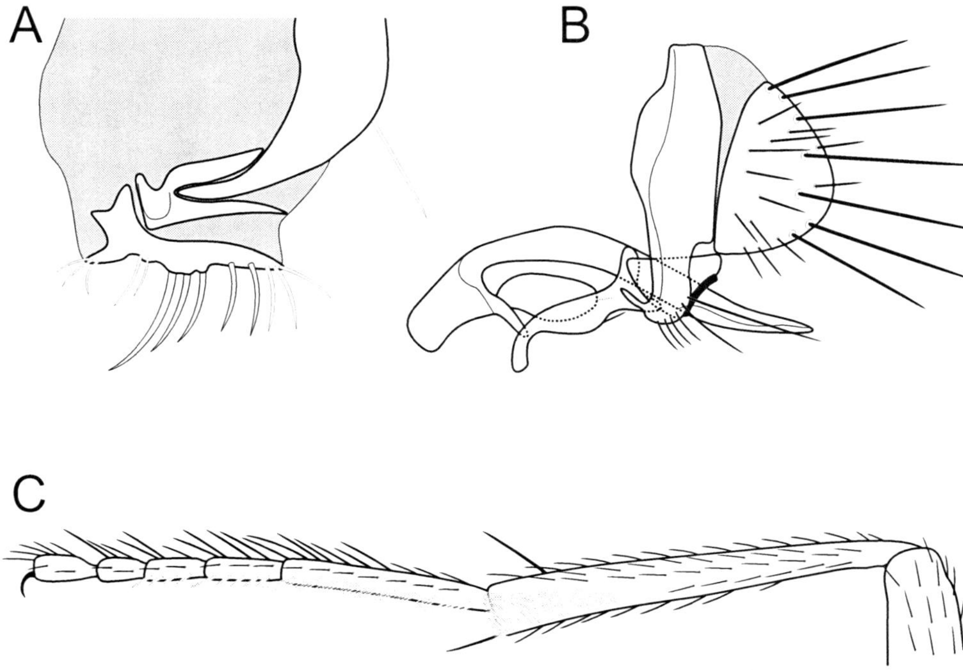

Fig. 6. *Drosophila komohana* male. (A) Lateral view of labellum. (B) Terminalia. (C) Right front leg, anterior view.

slightly infuscate anterior of R_{4+5} and with a faint mark over the dm–cu crossvein. Costal fringe extending about 2/5 the distance between the apex of R_{2+3} and R_{4+5}. **Abdomen.** Entirely dark brown. Epandrium narrowed dorsally; anteroventral spine curled under, thus appearing narrow in lateral view. Aedeagal apodeme longer than high, concave ventrally. Aedeagus narrow at the base, strongly bent, basal angle acute, apex elongate; preapical protuberance absent (Fig. 6B).
♀. Unknown.
MEASUREMENTS. N = 1♂. TL = 1.37 mm; WL = 2.94 mm; TL/WL = 0.5; HW = 1.13 mm; HW/FS = 2.3; HW/TL = 0.8; CI = 4.5; 4V = 1.5; 5X = 1.4; 4C = 0.5; M = 0.5.

TYPE. MAUI: Holotype ♂ (BPBM 16713), Pu'u Kukui, 4300 ft., Q30B, 20–21.vii.1970, KYK.

DISTRIBUTION & ECOLOGY. Maui; known only from the holotype. Breeding habits unknown.

ETYMOLOGY. From the Hawaiian *komohana*, west, referring its status as the only unplaced species collected from West Maui.

Drosophila lelolua **n. sp.**
Fig. 7

DIAGNOSIS. The presence of both dorsally and ventrally directed, prostrate, spinose setae on the labellum separates this species from all others.

DESCRIPTION. ♂. **Head.** Front brown, pale anteriorly; fronto-orbital plates and ocellar triangle darker and conspicuously pollinose. Anterior reclinate seta slightly posterior of the proclinate, about 2/3 as long; posterior reclinate about twice as long as the anterior. Ocellar setae slightly longer than the posterior reclinate. Vertical setae normal in position, about as long as the posterior reclinate. Eyes with relatively elongate pile, about as long as two facets. Face brown, with a broad carina medially. Antenna brown; arista with 5 dorsal and 2 ventral rays in addition to the apical fork. No oral vibrissae, all oral setulae similar. Gena yellow, narrowly brown along the oral margin. Palp yellow, flattened, only slightly expanded toward the apex, apex broadly rounded, setulose; with 2 elongate apical setae. Labellum with a short dorsal projection bearing 3 yellow, hair-like, medially-directed setae, the remainder of the labellum bearing the following setae, from dorsal to ventral (all are straight and prostrate): 3 straight, yellow, hair-like, dorsally-directed setae; 1 stout, brown, spine-like seta; a short gap, followed by 4 similar spine-like setae, which are ventrally-directed; and 3 brown hair-like setae, also ventrally-directed (Fig. 7A). Setae ventral of the gap are placed along the basal margin of the labellar sclerite, rather than the apical margin as usually seen. **Thorax.** Brown, pollinose; katepisternum somewhat paler. Two humeral setae, the ventral about 2/3 as long as the dorsal. Acrostichal setulae in 6–8 irregular rows. Two pairs of dorsocentral setae, the anterior about 2/3 as long as the posterior. Halteres brown. **Legs.** Yellow, femora faintly tinged with brown. Front tibia lacking elongate ciliation, although some dorsal setulae are longer than normal. Front tarsus with an anterodorsal and a posterior row of elongate cilia, each with 4 on the basitarsus, 2 on the second segment, and 1 on the third; fourth segment with a single dorsal cilia (Fig. 7C). **Wings.** Entirely hyaline. The dm–cu crossvein almost vertical in orientation and located more basally on the wing than in other *modified mouthparts* species. Costal fringe extending about halfway between the apex of R_{2+3} and R_{4+5}. **Abdomen.** Entirely brown. Epandrium slightly narrowed dorsally; anteroventral spine curled under, thus appearing narrow in lateral view. Aedeagal apodeme slightly higher than long, concave ventrally. Aedeagus narrow at the base, strongly bent, basal angle acute, apex broad and blunt; preapical protuberance indistinct (Fig. 7B).
♀. Unknown.
MEASUREMENTS. N = 1♂. TL = 1.10 mm; WL = 2.30 mm; TL/WL = 0.5; HW = 0.88 mm; HW/FS = 2.3; HW/TL = 0.8; CI = 3.8; 4V = 2.1; 5X = 1.9; 4C = 0.8; M = 0.7.

TYPE. LĀNA‘I: Holotype ♂ (BPBM 16717), Lāna‘i, 2400 ft., ex live *Freycinetia* leaf, R91, 11.iii.1973, SLM.

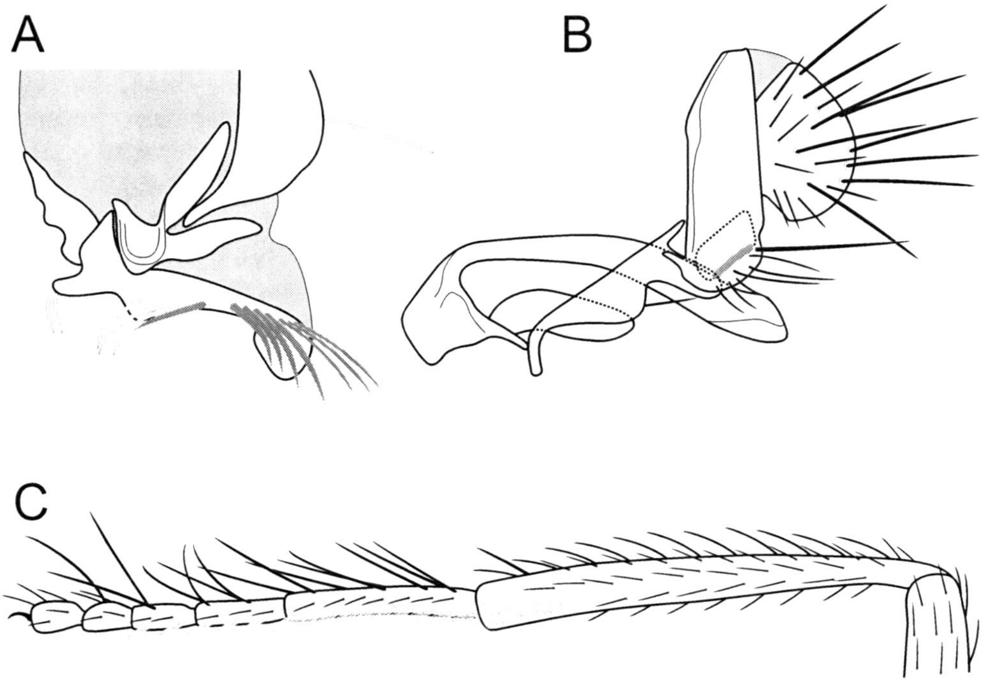

Fig. 7. *Drosophila lelolua* male. (A) Lateral view of labellum. (B) Terminalia. (C) Right front leg, anterior view.

Distribution & Ecology. Lāna'i. Known only from the holotype. Probably inhabits the same niche as several species in the *dissita* and *freycinetiae* subgroups, larvae feeding on decaying material that accumulates in the leaf axils of *Freycinetia* and *Pleomele*.

Etymology. From the Hawaiian *lelo* (tongue) + *lua* (double), hence double-tongued; a figurative expression, used here in the literal sense in reference to the opposing orientations of the labellar spines.

Drosophila omnivora n. sp.
Fig. 8A–C

Diagnosis. The quadrate labellar sclerite with only a few, widely-spaced spines somewhat resembles some members of the *fuscoamoeba* subgroup, but the wings are almost completely hyaline, infuscated only over the dm–cu crossvein, and the apical segment of vein CuA_1 is short, about as long as the dm–cu crossvein. The front basitarsi are also elongate, longer than the remaining tarsal segments combined. Very similar to *D. wahihuna*, separated by the form of the aedeagus.

DESCRIPTION. ♂. **Head.** Front rufous, paler anteriorly; fronto-orbital plates and ocellar triangle darker and conspicuously pollinose. Anterior reclinate seta about even with the proclinate, about 1/2–2/3 as long; posterior reclinate about 4 times as long as the anterior. Ocellar setae about as long as the posterior reclinate. Vertical setae normal in position, about as long as the posterior reclinate. Eyes with short, inconspicuous pile, about as long as one facet. Face yellow, with a rounded carina medially. Antenna pale brown; arista with 6–7 dorsal and 3 ventral rays in addition to the apical fork. Vibrissal angle with a row of 4–7 elongate setae, and another row of shorter setulae (similar to and continuous with the oral setulae) posterolaterally of the first. Gena yellow, narrowly brown along the oral margin. Palp yellow, flattened, weakly clavate, broadest near the apex, sparsely setulose, medial margin slightly concave; with an elongate subapical seta. Basal and apical margins of the labellar sclerite nearly parallel, not converging ventrally; sclerite also bearing a narrow, curved process from the ventral apex which curves around the ventral portion of the labellum. Labellum with 3 slightly curved, yellow, spine-like setae, of similar size and diameter; scattered hair-like setae also present (Fig. 8A). **Thorax.** Brownish yellow dorsally (possibly rufous to pale brown in fresh specimens), pollinose; pleura yellow with two brown stripes along the upper and lower margins of the anepisternum. Two humeral setae, the ventral about 3/4 as long as the dorsal. Acrostichal setulae in 6–8 irregular rows. Two pairs of dorsocentral setae, the anterior about 3/4 as long as the posterior. Halteres yellow, stems brownish. **Legs.** Yellow. Front tibia lacking ciliation. Front basitarsus with about 8 anterodorsal cilia in two irregular rows along the apical 2/3, mostly less than twice as long as the basitarsal diameter; second tarsal segment with 2 slightly longer cilia (Fig. 8C). **Wings.** Hyaline, faintly infuscated over the dm–cu crossvein. Costal fringe extending about halfway between the apex of R_{2+3} and R_{4+5}. Apical segment of vein CuA_1 short, about as long as the dm–cu crossvein; 5X ratio 0.8–1.1. **Abdomen.** Entirely light brown. Epandrium parallel-sided on the ventral half, strongly narrowed dorsally. Aedeagal apodeme rounded, ventral margin convex; about as long as high. Aedeagus narrow at the base, evenly curved and fishhook-shaped, apex narrow and elongate; preapical protuberance present, relatively small and rounded (Fig. 8B).

♀. Identical to the male with the following exceptions. **Head.** Labellum unmodified. **Legs.** Front legs without elongate cilia or setae. **Abdomen.** Ovipositor yellow, broadly triangular, apex blunt.

MEASUREMENTS. N = 5♂. TL = 1.48 (1.41–1.55) mm; WL = 2.98 (2.83–3.13) mm; TL/WL = 0.5; HW = 1.04 (0.95–1.10) mm; HW/FS = 2.2 (2.2–2.3); HW/TL = 0.7; CI = 4.8 (4.2–5.3); 4V = 1.2 (1.1–1.3); 5X = 1 (0.8–1.1); 4C = 0.4 (0.4–0.5); M = 0.3. N = 5♀. TL = 1.58 (1.45–1.70) mm; WL = 3.14 (2.95–3.23) mm; TL/WL = 0.5; HW = 1.05 (0.93–1.10) mm; HW/FS = 2.2 (2.1–2.2); HW/TL = 0.7 (0.6–0.7); CI = 5.0 (4.5–5.5); 4V = 1.2 (1.1–1.4); 5X = 1.0 (0.9–1.1); 4C = 0.5 (0.4–0.5); M = 0.3 (0.3–0.4).

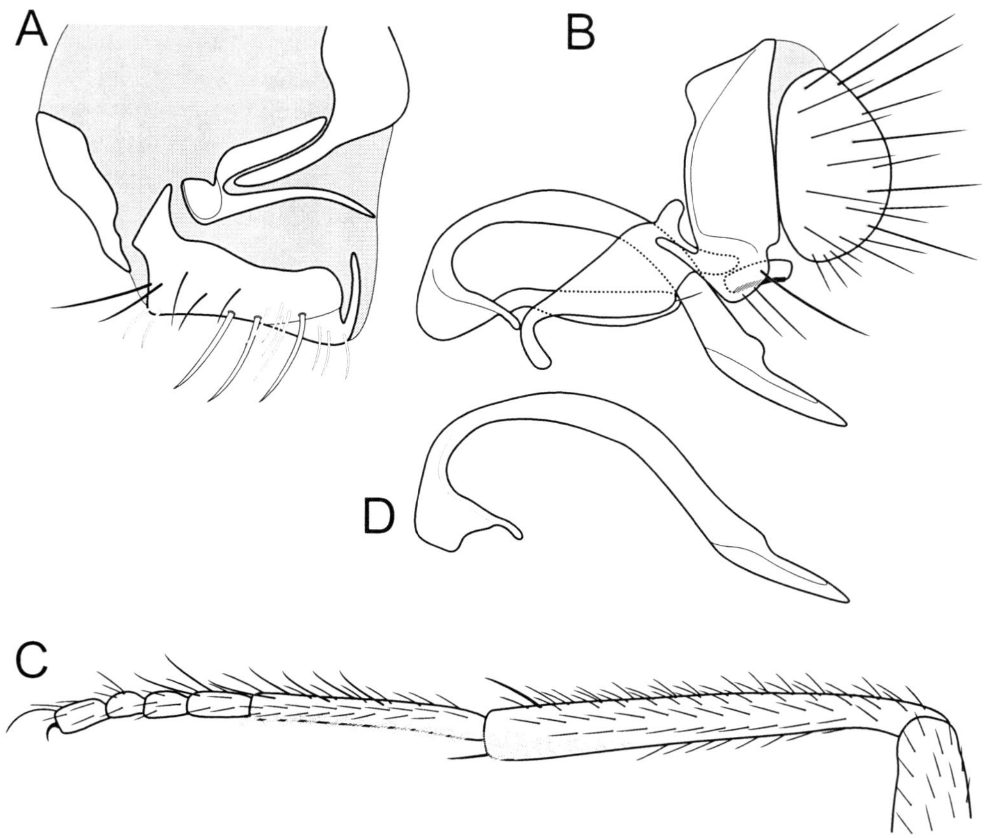

Fig. 8. *Drosophila omnivora* male. (A) Lateral view of labellum. (B) Terminalia. (C) Right front leg, anterior view. (D) *Drosophila wahihuna* male, aedeagus.

TYPES. O'AHU: Holotype ♂ (BPBM 16718) and allotype ♀, Kalua'a Gulch, Pu'u Hapapa, reared ex rotten fern frond, 15.ii.1970, SLM.

PARATYPES. 1♂, Mt. Ka'ala, "*Gunnera* spe.", 3600 ft., 22.vi.1939, FXW [paratype of *D. bridwelli*]. 1♂, Mt. Tantalus, C97.7, 18.vi.1964, HLC. 1♂, Palikea, J3, 15.vii.1966, WBH. 6♂ 1♀, same data as holotype; 2♂ 2♀, same data, reared ex rotten *Urera* stem. 8♂, Kalua'a Gulch, 1800–2100 ft., ex *Strongylodon* pod, P71, 21.ii.1971, SLM. 4♂ 5♀, Hidden Valley, ex *Touchardia* bark, Q23, 21.xi.1971, SLM (all UHIM).

DISTRIBUTION & ECOLOGY. O'ahu; most records from the Wai'anae range. Reared from a variety of hosts and substrates: *Touchardia latifolia* and *Urera* sp. bark (*olonā* and *ōpuhe*, Urticaceae), *Strongylodon ruber* pods, (*nuku 'i'iwi*, Fabaceae) and unidentified fern fronds (Magnacca et al. 2008).

ETYMOLOGY. From the Latin *omni* (all) + *vor* (eat), referring to the range of hosts used.

DISCUSSION. The rearing of this species from ferns, which are rarely used by Hawaiian drosophilids, suggests a possible relationship with the *fuscoamoeba* or *semifuscata* subgroups. This species and *D. wahihuna* are described as separate species because they are found on relatively distant islands and the morphological difference between them, while subtle, appears to be consistent. These are the two species mentioned by Magnacca and O'Grady (2007) as having mouthparts similar to *D. fuscoamoeba* but lacking wing marks and a sinuate R_{4+5} vein.

Drosophila tetraspilota Hardy

Drosophila tetraspilota Hardy, 1965:483–484. Holotype ♀, Upper Hāmākua Ditch Trail, 30.vii.1921, OHS, BPBM 6454 [examined].

DIAGNOSIS. An unmistakable species, easily recognized by the bright yellow thorax with strongly contrasting brown marks, and the wings with infuscation along the anterior margin and CuA_1.

DESCRIPTION. ♀. **Head.** Front yellow; fronto-orbital plates and ocellar triangle dark brown to black. Anterior reclinate seta placed even with the proclinate. Vertical setae normal in position. Eyes with short, inconspicuous pile, about as long as one facet. Face predominantly yellow-white, dark brown to black on the rounded median carina. Antenna brown, first two segments tinged with yellow; arista with 8 dorsal and 2 ventral rays in addition to the apical fork. One strong oral vibrissa present. Gena yellow. Palp dark brown to black, paler at the base, flattened, sparsely setulose; one strong apical seta present. Labellum without modifications. **Thorax.** Mesonotum largely pale yellow, with two submedian longitudinal brown stripes extending from just anterior of the anterior dorsocentral to the posterior margin, fading posteriorly, and a brown spot on each side near the wing base. Humeri brown. Pleura yellow, extensively marked with brown: a small spot behind the humerus, a median stripe through the anepisternum, and median spots on the katepimeron and metapleuron; katepisternum almost entirely brown. Disc of scutellum brown, yellow on the edges; metanotum brown. **Legs.** Entirely yellow. Front legs without elongate cilia or setae. **Wings.** Infuscated across the entire anterior margin and along the apical 3/4 of vein CuA_1, the latter mark more diffuse. around the dm–cu crossvein and along the entire costal margin, broadening and becoming more diffuse around the wing apex, extending to the apex of M. Costal fringe extending about half the distance between the apex of R_{2+3} and R_{4+5}. **Abdomen.** First segment yellow, remainder reddish brown. Ovipositor broadly triangular, apex rounded.
♂. Unknown.

MATERIAL EXAMINED. 1♀, Upper Waiākea Forest Reserve, R Road kīpuka, 3200 ft., sweeping *Pritchardia*, 14.viii.2005, KNM (UCPO).

DISTRIBUTION & ECOLOGY. Hawai'i; extremely rare. Breeding habits unknown.

DISCUSSION. The female mesonotum and wing were illustrated by Hardy (1965, Fig. 39). The coloration of the thorax and wing are unique and make this species extremely difficult to place without the male. Based on the ovipositor, it appears to be related to the *modified mouthparts* group. The only Hawaiian *Drosophila* species with a similar wing pattern is *D. truncipenna*, which is a member of the *adiastola* subgroup of the *picture wing* group. However, females of the *adiastola* subgroup have relatively short, triangular ovipositors rather than the narrow, elongate form found in other *picture wing* species and some males (e.g. *D. ornata*) have spinose mouthparts. As a result, it has been suggested that the *adiastola* subgroup may be more closely related to the *modified mouthparts* group than to the other *picture wing* subgroups (Kaneshiro 1997).

Drosophila toxacantha n. sp.

Fig. 9

DIAGNOSIS. Differs from all other species by having the dorsal labellar spine black, elongate, and curved, and the remaining two much shorter and straight; and by having only posteroventral cilia on the front tarsus.

DESCRIPTION. ♂. **Head.** Front brown, paler anteriorly; fronto-orbital plates and ocellar triangle slightly darker. Anterior reclinate seta slightly posterior of the proclinate, about equal in length; posterior reclinate about twice as long. Ocellar setae slightly longer than the posterior reclinate. Vertical setae normal in position, about as long as the posterior reclinate. Eyes with short, inconspicuous pile, about as long as one facet. Face yellow, with a rounded carina medially. Antenna pale brown; arista with 3 dorsal and 2 ventral rays in addition to the apical fork. One strong oral vibrissa, the next seta 1/2–2/3 as long, remaining oral setulae smaller. Gena yellow, narrowly brown along the oral margin. Palp yellow, flattened, clavate, broadest at about 2/3 the length, sparsely setulose; one strong apical seta. Labellum with three dorsal spines: the first is black, elongate, and curved dorsally; the second is black, short, and nearly straight; and the third is brown and curved medially (Fig. 9A). The ventral portion of the labellum is covered with about 12 black hair–like setae. **Thorax.** Almost entirely brown, pollinose; with a faint yellow stripe along the anapleural suture. Two humeral setae, the ventral about 3/4 as long as the dorsal. Acrostichal setulae in 6 rows. Two pairs of dorsocentral setae, the anterior about 3/4 as long as the posterior. Halteres yellow. **Legs.** Yellow. Front tibia with one elongate posteroventral cilia, about 1/3 as long as the tibia. Front basitarsus with 4 posteroventral cilia (Fig. 9C). **Wings.** Entirely hyaline. Costal fringe extending 1/4 or less of the distance between the apex of R_{2+3} and R_{4+5}. **Abdomen.** Brown, paler on the basal half of each segment. Cerci darker and enlarged, nearly semicircular, about 2/3 as long as high. Epandrium short, parallel-sided. Aedeagal

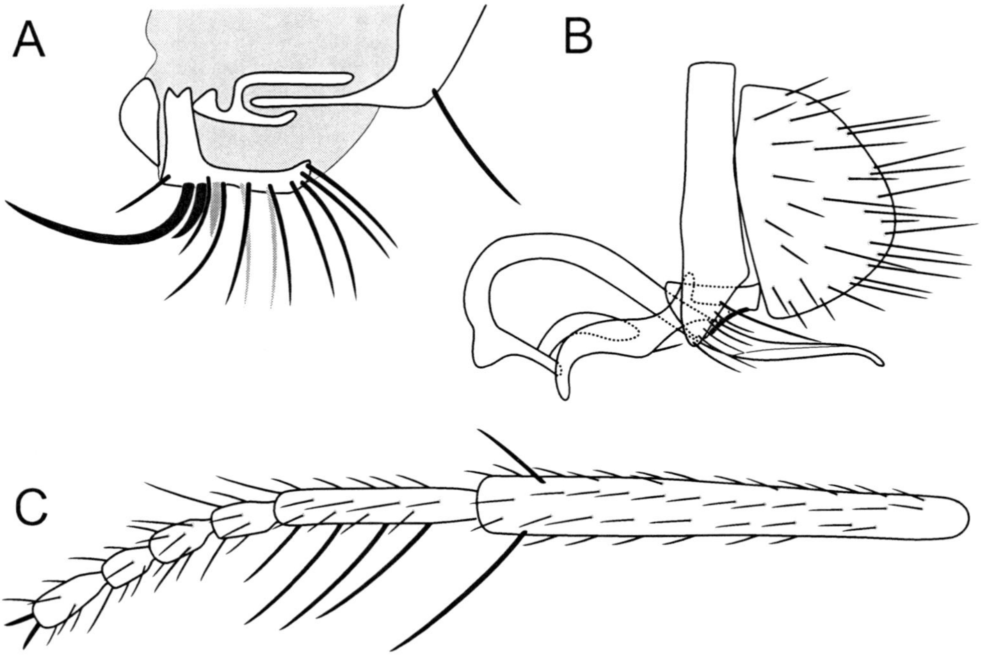

Fig. 9. *Drosophila toxacantha* male. (A) Lateral view of labellum. (B) Terminalia. (C) Left front leg, posterodorsal view.

apodeme very small and rounded. Aedeagus evenly curved and fishhook-shaped, apex thin and elongate; preapical protuberance absent (Fig. 9B).

♀. Identical to the male with the following exceptions. **Head.** Labellum unmodified. **Legs.** Front legs without elongate cilia or setae. **Abdomen.** Ovipositor yellow, broadly triangular, apex blunt.

MEASUREMENTS. N = 5♂. TL = 0.86 (0.83–0.93) mm; WL = 1.86 (1.80–1.98) mm; TL/WL = 0.5; HW = 0.66 (0.63–0.70) mm; HW/FS = 2.4 (2.3–2.6); HW/TL = 0.8; CI = 3.3 (3.0–3.5); 4V = 1.8 (1.8–1.9); 5X = 2.2 (2.0–2.6); 4C = 0.8 (0.7–0.9); M = 0.6 (0.6–0.7). N = 1♀. TL = 1.03 mm; WL = 2.23 mm; TL/WL = 0.5; HW = 0.70 mm; HW/FS = 2.2; HW/TL = 0.7; CI = 3.6; 4V = 1.8; 5X = 2.2; 4C = 0.7; M = 0.7.

TYPES. HAWAI'I: Holotype ♂ (BPBM 16709) and allotype ♀, Pīhā–Maulua, 6500 ft., 12.vii.1967, reared ex *Cheirodendron* branches, K14, WBH.

PARATYPES. 5♂, same data as holotype (UHIM).

OTHER MATERIAL. 1♂, Kūlani Cone, reared ex *Cheirodendron* branches, 14.i.1999, KH (HVIC). 3♂, Ho'okena, Kona Forest Unit of Hakalau NWR, 5500 ft., 19.vi.2000, KH (HVIC).

DISTRIBUTION & ECOLOGY. Hawai'i; rarely collected but apparently widespread in wet forest. Reared from bark of *Cheirodendron trigynum* ('ōlapa, Araliaceae).

ETYMOLOGY. From the Greek *toxon* (bow) + *acantha* (spine or thorn), referring to the curved dorsal spine of the labellum.

DISCUSSION. This species was referred to as "n. sp. nr. *latigena*" in Heed (1968). Although the presence of black spines implies a connection with the *mitchelli* subgroup, mtDNA sequences show it to be much closer to the *dissita/mimica* complex (unpubl. data).

Drosophila umiumi n. sp.

Fig. 3E

DIAGNOSIS. See *D. barbata*.

DESCRIPTION. Identical to *D. barbata* except slightly darker overall, yellow stripe on pleura indistinct, and front legs of male without long cilia (Fig. 3E).
MEASUREMENTS. N = 3♂. TL = 1.27 (1.18–1.38) mm; WL = 2.71 (2.60–2.79) mm; TL/WL = 0.5 (0.4–0.5); HW = 0.94 (0.85–1.01) mm; HW/FS = 2.0 (2.0–2.1); HW/TL = 0.7 (0.7–0.8); CI = 4 (3.8–4.4); 4V = 1.6 (1.5–1.6); 5X = 1.7 (1.6–1.8); 4C = 0.6; M = 0.5. N = 3♀. TL = 1.57 (1.53–1.64) mm; WL = 3.18 (3.08–3.32) mm; TL/WL = 0.5; HW = 1.08 (1.05–1.13) mm; HW/FS = 2.1 (2.0–2.1); HW/TL = 0.7; CI = 4.7 (4.6–4.8); 4V = 1.6 (1.6–1.7); 5X = 1.6 (1.5–1.8); 4C = 0.5 (0.5–0.6); M = 0.5 (0.5–0.6).

TYPES. HAWAI'I: Holotype ♂ (BPBM 16719) and allotype ♀, Kahuku Ranch, ex bark of *Clermontia*, U50B, 1–3.ii.1978, KYK.

PARATYPES. 2♀, same data as holotype. 1♂, Pauahi, 4500 ft., ex *Clermontia* stem, 3.iv.1972, KYK. 1♂ Kahuku Ranch, 4050 ft., T69, 14.viii.1975, KYK (all UHIM).

DISTRIBUTION & ECOLOGY. West Hawai'i: Ka'ū and South Kona districts. Probably occurs in North Kona as well. Reared from bark and stems of *Clermontia* spp.; probably has a host range similar to *D. barbata*.

ETYMOLOGY. From the Hawaiian *'umi'umi*, whiskers or beard, referring to the setation of the mouthparts and similarity to *D. barbata*.

DISCUSSION. The paratype from collection T69 appears to be aberrant; it has two inner vertical setae on each side of the head, but otherwise matches the holotype. See Discussion of *D. barbata*.

Drosophila wahihuna n. sp.
Fig. 8D

DIAGNOSIS. See *D. omnivora.*

DESCRIPTION. Identical to *D. omnivora* except that the aedeagal apodeme is concave on the ventral margin rather than convex (Fig. 8D).
MEASUREMENTS. N = 3♂. TL = 1.5 (1.46–1.55) mm; WL = 2.88 (2.80–2.95) mm; TL/WL = 0.5; HW = 1.10 (1.04–1.15) mm; HW/FS = 2.2 (2.2–2.3); HW/TL = 0.7 (0.7–0.8); CI = 4.7 (4.6–4.8); 4V = 1.3; 5X = 1.0; 4C = 0.5; M = 0.3 (0.3–0.4). N = 2♀. TL = 1.38 (1.25–1.51) mm; WL = 2.69 (2.54–2.85) mm; TL/WL = 0.5; HW = 0.96 (0.88–1.04) mm; HW/FS = 2.2; HW/TL = 0.7; CI = 4.6; 4V = 1.2; 5X = 1; 4C = 0.5; M = 0.3.

TYPES. KAUA'I: Holotype ♂ (BPBM 16720) and allotype ♀, collection C53.13 [no other data].

PARATYPES. 2♂ 1♀, same data as holotype (UHIM).

DISTRIBUTION & ECOLOGY. Kaua'i; known only from the type series. Breeding habits unknown; probably with a similar host range to *D. omnivora.*

ETYMOLOGY. From the Hawaiian *wahi* (place) + *huna* (hidden or secret), from the lack of locality data on the specimen labels.

Drosophila waikamoi n. sp.
Fig. 10

DIAGNOSIS. See *D. komohana.*

DESCRIPTION. ♂. **Head.** Front brown, slightly paler and more pollinose on fronto-orbital plates and ocellar triangle. Anterior reclinate seta slightly posterior of the proclinate, about half as long; posterior reclinate about 3 times as long as the anterior. Ocellar setae about as long as the posterior reclinate. Vertical setae normal in position, about as long as the posterior reclinate. Eyes with short, inconspicuous pile, about as long as one facet. Face brown, with a rounded carina. Antenna dark brown, paler on the dorsolateral surface of the second segment; arista with 4 dorsal and 2 ventral rays in addition to the apical fork. Several elongate oral setae present at the vibrissal angle (about twice as long as the posterior oral setulae), but no distinct oral vibrissae. Gena yellow, narrowly brown along the oral margin. Palp yellow, flattened, broadest near the apex, setulose; with 2 elongate subapical setae along the lateral margin. Labellum laterally with 2 strong, brown, curved, round, spine-like setae, followed by 2–3 weaker spines, with a few inconspicuous hair-like setae at the dorsal and ventral ends (Fig 10A). **Thorax.** Brown and pollinose, except the katepisternum which is pale. Two humeral setae, the ventral about 2/3 as long as the dorsal. Acrostichal setulae in 6–8 irregular rows. Two

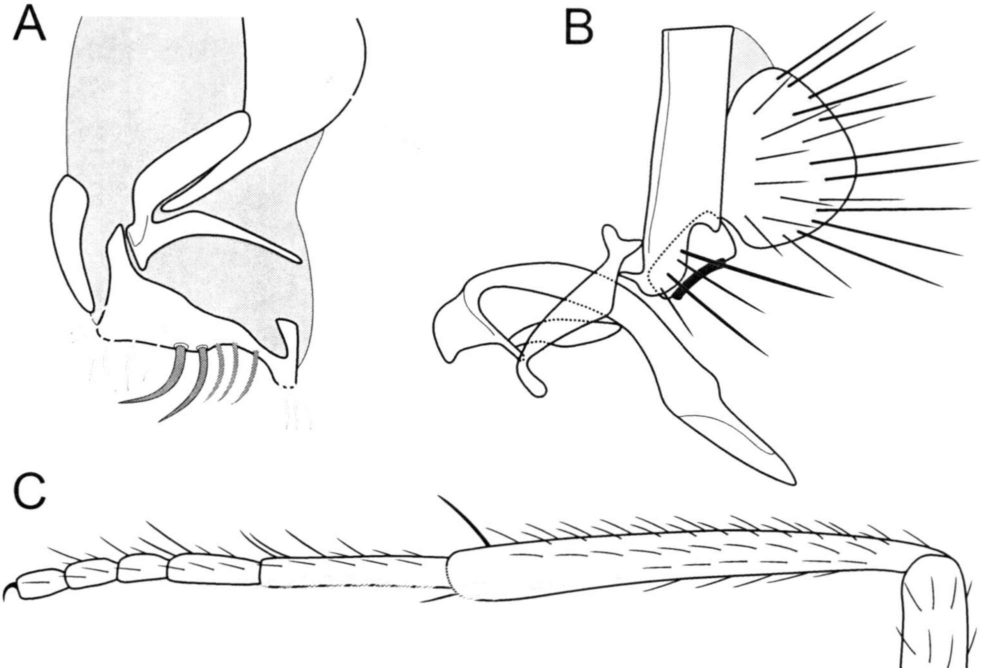

Fig. 10. *Drosophila waikamoi* male. (A) Lateral view of labellum. (B) Terminalia. (C) Right front leg, anterior view.

pairs of dorsocentral setae, the anterior about 2/3 as long as the posterior. Halteres pale brown. **Legs.** Yellow, tinged with brown. Front tibia lacking ciliation. Front basitarsus with 4 relatively short dorsal cilia on the apical half, remaining segments with 1–2 each (Fig. 10C). **Wings.** Mostly hyaline, slightly smoky adjacent to R veins and faintly marked on the dm–cu crossvein. Costal fringe probably extending about 2/5–1/2 the distance between the apex of R_{2+3} and R_{4+5} (wingtip broken in holotype). **Abdomen.** Entirely dark brown. Epandrium parallel-sided; anteroventral spine curled under, thus appearing narrow in lateral view. Aedeagal apodeme not much longer than high, concave ventrally. Aedeagus narrow at the base, strongly bent, basal angle acute, apex elongate; preapical protuberance small, low (Fig. 10B).

♀. Unknown.

MEASUREMENTS. N = 1♂. TL = 0.97 mm; WL = 2.87 mm (estimated, wing damaged); TL/WL = 0.3; HW = 0.81 mm; HW/FS = 2.1; HW/TL = 0.8; CI = 4.2; 4V = 1.5; 5X = 1.6; 4C = 0.5; M = 0.6.

TYPE. MAUI: Holotype ♂ (BPBM 16714), Waikamoi, Flume Trail, 14.viii.1964, DEH.

DISTRIBUTION & ECOLOGY. Maui; known only from the holotype. Breeding habits unknown.

ETYMOLOGY. From the type locality.

DISCUSSION. A single specimen of a similar species from O'ahu is in the UHIM, but it is in poor condition.

Drosophila wikstroemiae n. sp.
Fig. 11

DIAGNOSIS. Resembles *D. barbata* and some members of the *hirtitarsus* and *semifuscata* groups in having weak labellar spines and elongate tarsal cilia, and with an elongate front basitarsus as in *D. omnivora*. Differing from all of these by lacking cilia on the third segment of the front tarsus.

DESCRIPTION. ♂. **Head.** Front brown, pale anteriorly; fronto-orbital plates and ocellar triangle slightly darker and more pollinose. Anterior reclinate seta slightly posterior of the proclinate, about 2/3 as long; posterior reclinate about 2.5–3 times as long as the anterior. Ocellar setae about as long as the posterior reclinate. Vertical setae normal in position, about as long as the posterior reclinate. Eyes with short, inconspicuous pile, about as long as one facet. Face brown, weakly with a rounded carina medially. Antenna brown, pale near the junction of the second and third segments; arista with 5–6 dorsal and 3 ventral rays in addition to the apical fork. Two moderately strong oral vibrissae present. Gena yellow, narrowly brown along the oral margin. Palp yellow, flattened, broadest at about 2/3 its length, setulose; with a strong apical seta. Labellum laterally with 2 thin, brown, curved, round, spine-like setae, followed by 5 hair-like setae; additional scattered hair-like setae present dorsally and ventrally (Fig. 11A). **Thorax.** Mesonotum and upper half of pleura brown, pollinose; katepisternum yellow. Two humeral setae, the ventral nearly as long as the dorsal. Acrostichal setulae in 6–8 irregular rows. Two pairs of dorsocentral setae, the anterior about 2/3 as long as the posterior. Halteres yellow, stems brownish. **Legs.** Yellow, tinged faintly with brown. Front tibia lacking ciliation. Front basitarsus with 4 elongate anterodorsal cilia along the apical half; second segment with 3 similar cilia; both segments with an elongate posterodorsal cilia near the apex (Fig. 11C). **Wings.** Hyaline, faintly infuscated over the dm–cu crossvein. Costal fringe extending about halfway between the apex of R_{2+3} and R_{4+5}. **Abdomen.** Brown, each segment slightly paler anterolaterally. Epandrium slightly broader along the dorsal margin than at the ventral lobe, anteroventrally with a broad spine. Aedeagal apodeme rounded, semicordate. Aedeagus narrow at the base, evenly curved and fishhook-shaped, apex relatively short; preapical protuberance prominent (Fig. 11B).
♀. Identical to the male with the following exceptions. **Head.** Three oral vibrissae present. Labellum unmodified. **Legs.** Front legs without elongate cilia or setae. **Abdomen.** Ovipositor yellow, broadly triangular, apex blunt.

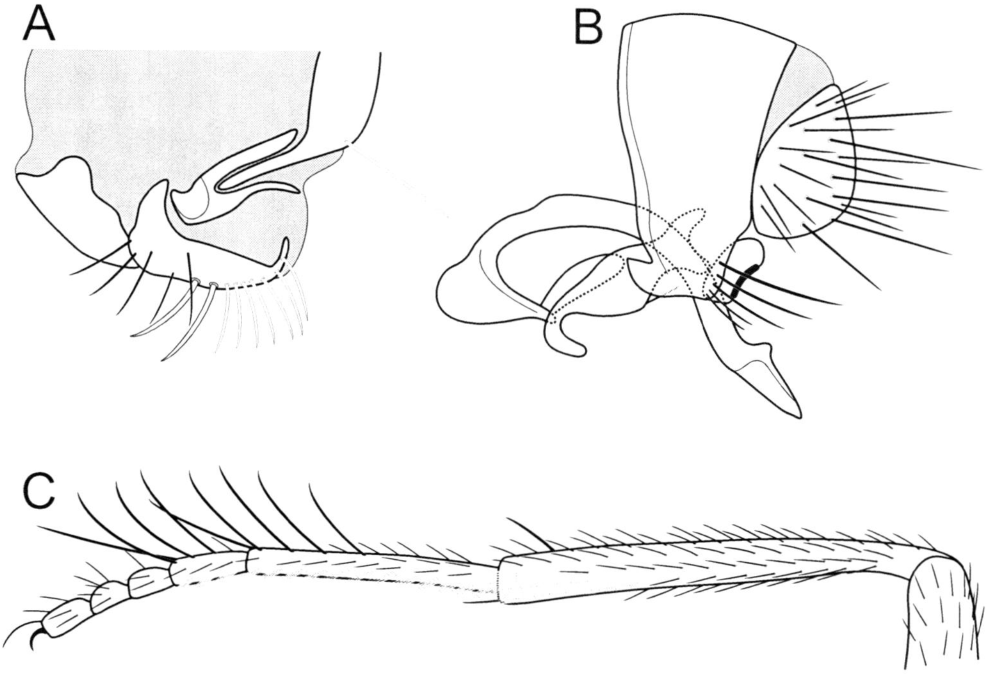

Fig. 11. *Drosophila wikstroemiae* male. (A) Lateral view of labellum. (B) Terminalia. (C) Right front leg, anterior view.

MEASUREMENTS. N = 3♂. TL = 1.21 (1.15–1.25) mm; WL = 2.51 (2.42–2.59) mm; TL/WL = 0.5; HW = 0.93 (0.87–0.99) mm; HW/FS = 2.3 (2.3–2.4); HW/TL = 0.8; CI = 4.1 (3.9–4.4); 4V = 1.5 (1.5–1.6); 5X = 1.8 (1.8–2); 4C = 0.6; M = 0.5. N = 1♀. TL = 1.31 mm; WL = 2.64 mm; TL/WL = 0.5; HW = 0.95 mm; HW/FS = 2.3; HW/TL = 0.7; CI = 3.7; 4V = 1.6; 5X = 1.5; 4C = 0.6; M = 0.5.

TYPES. O‘AHU: Holotype ♂ (BPBM 16721) and allotype ♀, Ka‘au Crater rim, 1600 ft., ex rotting *Wikstroemia*, 29.vi.1987, ATO & WDP.

PARATYPES. 2♂, same data as holotype. 6♂, Wai‘anae Mts., P7Q, 22.ii.1970, SLM (all UHIM).

DISTRIBUTION & ECOLOGY. O‘ahu. Reared from bark of *Wikstroemia* (‘ākia, Thymeliaceae).

ETYMOLOGY. From the host plant.

DISCUSSION. Because the labellar spines are rather thin and may not be distinguishable from the hair-like setae, this species is keyed out twice. The mouthparts are superficially similar to the *hirtitarsus* subgroup in lacking strong spines. However, based on the form of the aedeagus and epandrium, *D. wikstroemiae* appears to be more closely related to the *freycinetiae* subgroup or to *D. barbata*.

ceratostoma subgroup

1. Appendage of labellum unbranched, with few small, black setae (Fig. 13A). Hawai'i .... *D. wikani* n. sp.
— Appendage divided into two or three branches, with flattened, pale, scale-like setae .... 2.

2. (1) Appendage with a conspicuous brush of long, scale-like setae all around the apex (Fig. 12A). Hawai'i .... *D. orascopa* n. sp.
— Appendage with some scale-like setae but not covering the entire apex...... .... 3.

3. (2) Femora dark brown; front basitarsus laterally compressed and expanded, with dense, conspicuous cilia on the anterior surface. Kaua'i.................... .... *D. humeralis* Grimshaw
— Femora yellow; front basitarsus not noticeably flattened or expanded, cilia anterodorsal and relatively inconspicuous. Hawai'i .... *D. ceratostoma* Hardy

Drosophila ceratostoma Hardy

Drosophila ceratostoma Hardy, 1966:203–205. Holotype ♂ and allotype ♀, Kīpuka Kī, 12.ix.1964, HTS. BPBM 11259 [examined].

Diagnosis. Similar to *D. humeralis* and *D. orascopa*, distinguished by the shorter, multifurcate labellar process and pale coloration of the pleura and legs.

Description. ♂. **Head.** Front brown, paler anteriorly; fronto-orbital plates and ocellar triangle darker but conspicuously pollinose, appearing gray. Anterior reclinate seta even with the proclinate, about 2/3 as long; posterior reclinate about 3 times as long as the anterior. Ocellar setae about as long as the posterior reclinate. Vertical setae normal in position, the inner about 3/4 as long as the outer. Eyes with short, inconspicuous pile, about as long as one facet. Face yellow, with a rounded carina medially. The eyes are strongly convergent below, so that the lower portion of the face is narrowed. Antenna brown; arista with 4–5 dorsal and 2 ventral rays in addition to the apical fork. No strong oral vibrissae. Gena yellow. Palp yellow, narrow, flattened, sparsely setulose; one moderately strong, short apical seta. Fleshy portion of labellum absent. Labellar sclerite highly modified into a sclerotized process: the basal half is quadrate, from which emerge a short, narrow ventral branch and a longer, broader dorsal branch. The latter splits again near its apex into three branches; the ventralmost of these bears several elongate, pale, flattened setae, while the other two have elongate, black, pointed setae (it is unclear if the latter consist of a single seta each or several appressed together). **Thorax.** Brown, pollinose; paler on the humeri and margins of pleural sclerites. One humeral seta.

Acrostichal setulae in 6–8 irregular rows. Two pairs of dorsocentral setae, the anterior about 3/5 as long as the posterior. Halteres yellow. **Legs.** Yellow. Front tibia lacking ciliation. Front tarsus with one row each of short, inconspicuous anterodorsal and posterodorsal cilia, these only about as long as the width of the segment; basitarsus also with a similar ventral row of cilia; segments 3 and 4 with 1–2 stronger dorsal cilia, about twice as long as the others. **Wings.** Faintly infuscated around the dm–cu crossvein and anteroapically along R_{2+3}, R_{4+5}, and M. Costal fringe extending about 1/3 the distance between the apex of R_{2+3} and R_{4+5}. **Abdomen.** Dark brown. Genitalia not examined; illustrated by Hardy (1966).
♀. Identical to the male with the following exceptions. **Head.** Inner vertical seta slightly longer than the outer. Face not narrowed ventrally. One strong oral vibrissa. Labellum unmodified. **Legs.** Front legs without elongate cilia or setae. **Abdomen.** Ovipositor elongate triangular, apex rounded.

MATERIAL EXAMINED. HAWAI'I: 1♂ 1♀ paratypes, same data as holotype (UHIM).

DISTRIBUTION & ECOLOGY. Hawai'i; extremely rare, known only from the type collection.

DISCUSSION. The male mouthparts, genitalia, front leg, and wing, and ovipositor were illustrated by Hardy (1966, Fig. 4a, c-f). Hardy (1966) describes the type collection as "reared from bracket fungus," but one specimen has "reared from" crossed out and "feeding on" handwritten in, while two others have no label regarding how they were collected. Feeding seems more likely, since many Hawaiian *Drosophila* are attracted to rotting fungi but only a few outside the *haleakalae* group breed on them. In any event, it is likely that this species is the first one named as "n. sp. rel. *hirtitarsus*" by Heed (1968), although the date for that collection is given as Sept. 10, 1964. Despite the fact that Kīpuka Kī and nearby Kīpuka Puaulu are among the most intensively sampled sites for *Drosophila*, it has never been recollected.

Drosophila humeralis Grimshaw

Drosophila humeralis Grimshaw, 1901:64–65. Type ♂ and ♀, Kaua'i, 4000 ft., vii.1894, RCLP. BMNH [examined].

Drosophila aethostoma Hardy and Kaneshiro, 1968:247–250. Holotype ♂ and allotype ♀, Alaka'i Swamp, 4000 ft., G84B, 22.vi.1966, KYK. BPBM 8925 [examined].

DIAGNOSIS. Readily separated from other members of the subgroup by the Y-shaped labellar process, which lacks conspicuous scale-like setae; the long, thin inner vertical bristle; and the infuscation of the wing, which extends across the costal margin.

DESCRIPTION. ♂. **Head.** Front dark brown, grading into yellow anteriorly; fronto-orbital plates and ocellar triangle conspicuously pollinose, slightly grayish. Anterior reclinate seta placed slightly above the proclinate, 1/2–2/3 as long; posterior reclinate about 3 times as long as the anterior. Ocellar setae slightly longer than the posterior reclinate. Vertical setae normal in position, the inner about equal to the outer in length but much thinner and more curved. Eyes with short, inconspicuous pile, about as long as one facet. Face brown, with a rounded carina medially. Antenna brown, third segment darker, almost black; arista with 5–6 dorsal and 3 ventral rays in addition to the apical fork. No strong oral vibrissae. Gena yellow. Palp brown, paler at the base, flattened, sparsely setulose; one moderately strong, short apical seta and a longer subapical seta medially. Fleshy portion of labellum absent. Labellar sclerite highly modified into a sclerotized process, which splits into medial and lateral branches about halfway along its length. The medial branch bears a thick, black, flattened seta laterally and several thinner setae medially at its apex; the lateral branch has a short, thick, black, hooked seta and several thinner, brown, inclinate setae at the apex and 3–4 elongate setae along the apical half of the lateral margin. **Thorax.** Largely dark brown, densely pollinose; yellow on the humeri, in a stripe along the lower margin of the anepisternum, and along the rim of the scutellum. Central portion of mesonotum often depressed, flat or concave. Only one strong humeral seta; the ventral seta is reduced, only about 1/3 as long as the dorsal and hardly distinguishable from the surrounding setulae. Two pairs of dorsocentral setae, the anterior about half as long as the posterior. Acrostichal setulae in 12 or more irregular rows; those between the dorsocentrals elongate, over 3 times as long as the anterior acrostichals and nearly as long as the anterior dorsocentral, although much weaker. Halteres yellow, stems tinged with brown. **Legs.** Coxae and femora brown, tibiae and tarsi yellow. Front tibia lacking ciliation. Front basitarsus anteroposteriorly flattened and slightly expanded dorsally; posterior surface evenly covered with with short erect cilia, anterior face with 3 distinct rows of elongate cilia; also with a row of dorsal cilia along the apical half. Remaining segments of front tarsus each with 2–3 elongate dorsal cilia at their apices; also with a few anterior and posterior cilia corresponding to those on the basitarsus. **Wings.** Infuscated around the dm–cu crossvein and along the entire costal margin, broadening and becoming more diffuse around the wing apex, extending to the apex of M. Costal fringe extending about half the distance between the apex of R_{2+3} and R_{4+5}. **Abdomen.** Entirely dark brown. Epandrium slightly narrower dorsally than at the ventral margin. Aedeagal apodeme more or less quadrate, ventral margin concave; short, slightly higher than long. Aedeagus with basal angle acute; moderately arched beyond the apodeme, apex narrow and elongate; preapical protuberance present but small.

♀. Identical to the male with the following exceptions. **Head.** One strong oral vibrissa. Labellum unmodified. **Legs.** Front legs without elongate cilia or setae, tarsus not flattened. **Abdomen.** Ovipositor elongate triangular, apex rounded.

MATERIAL EXAMINED. Originally identified as *D. aethostoma*: KAUA'I: 1♂ paratype, same data as holotype. 1♂, Kōke'e, Plum Tree Trail, 3.viii.1968, HTS. 3♂ 1♀,

Kōke'e, Plum Tree Trail, 10.viii.1968, HTS. 2♂ 1♀, Powerline Trail, 2400 ft., 4.ix.1970, SLM. 4♂, Makaleha, 2500 ft., ex *Clermontia* stem, 8.ii.1981, SLM. Originally identified as *D. humeralis*: Kaua'i: 1♂ 1♀, Alakai Swamp, 4000 ft., HH20, reared from *Clermontia* fruit, 22.vii.1964, WBH (all above at UHIM). 1♂, Pihea Trail, 3900 ft., on and under *Clermontia*, 18.v.2007, KNM (UCPO). 2♂, Pihea Trail, 3900 ft., sweeping vegetation and ground, 18.v.2007, KNM; 1 ♀, same data, 4100 ft. (UCPO).

Distribution & Ecology. Kaua'i; uncommon but relatively widespread in wet forest. Reared from bark and fruit of *Clermontia* ('ōhā wai, Campanulaceae).

Discussion. The male head, mouthparts, genitalia, front leg, and wing, and ovipositor were illustrated by Hardy and Kaneshiro (1968, Fig. 32) in the description of *D. aethostoma*. The type male of *D. humeralis* has the distinctive mouthparts folded back and concealed, which probably explains why they were not noticed by either Grimshaw or Hardy. Though by no means abundant, this is the most frequently collected member of the *ceratostoma* subgroup, and the only one with reliable rearing records.

Drosophila orascopa n. sp.

Fig. 12

Diagnosis. The mouthparts of this species cannot be confused with any other, with the dense brush of elongate, scale-like setae.

Description. ♂. **Head.** Front brown, paler anteriorly; fronto-orbital plates and ocellar triangle conspicuously pollinose. Anterior reclinate seta slightly posterior of the proclinate, about half as long; posterior reclinate about 4 times as long as the anterior. Ocellar setae nearly as long as the posterior reclinate. Vertical setae normal in position, the inner about 3/4 as long as the outer. Eyes with short, inconspicuous pile, about as long as one facet. Face brown, with a rounded carina medially. Antenna brown; arista with 7 dorsal and 3 ventral rays in addition to the apical fork. No strong oral vibrissae. Gena yellow. Palp yellow, flattened, sparsely setulose; two strong apical setae. Labellum highly modified into an elongated sclerotized process; on each side it is forked about halfway down, and the apex of each fork is covered in elongate, pale, flattened setae (Fig. 12A). The dorsal fork also has two strong brown setae (may be black in life). **Thorax.** Entirely brown, pollinose. Two humeral setae, the ventral about 3/4 as long as the dorsal. Acrostichal setulae in 6–8 irregular rows. Two pairs of dorsocentral setae, the anterior about 3/5 as long as the posterior. Halteres pale brown. **Legs.** Yellow, tinged with brown. Front tibia lacking ciliation. Front tarsus with only short, inconspicuous cilia, hardly longer than the straight setulae (Fig. 12B). **Wings.** Entirely hyaline. Costal fringe extending about half the distance between the apex of R_{2+3} and R_{4+5}. **Abdomen.** Entirely dark brown. Genitalia not dissected.
♀. Unknown.

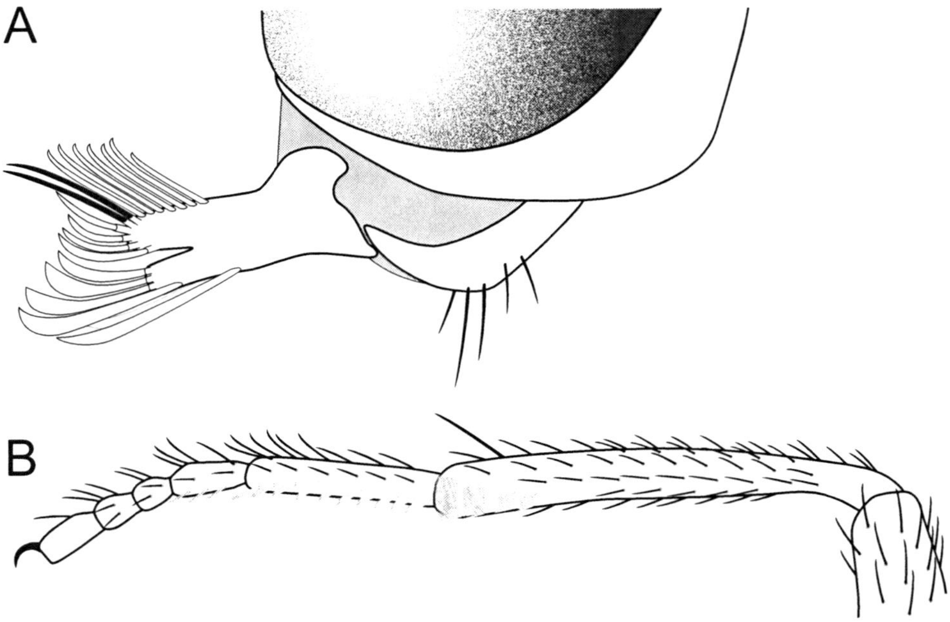

Fig. 12. *Drosophila orascopa* male. (A) Lateral view of labellum and lower head (drawn from pinned specimen). (B) Right front leg, anterior view.

MEASUREMENTS. N = 1♂. TL = 1.14 mm; WL = 2.30 mm; TL/WL = 0.5; HW = 0.90 mm; HW/FS = 2.3; HW/TL = 0.8; CI = 4.5; 4V = 1.6; 5X = 1.4; 4C = 0.5; M = 0.5.

TYPE. HAWAIʻI: Holotype ♂ (BPBM 16704), Kohala, Pololū Valley, M69, 21.viii.1969, KYK.

DISTRIBUTION & ECOLOGY. Hawaiʻi; known only from the holotype. Breeding habits unknown.

ETYMOLOGY. From the Latin *ora* (mouth) + *scopa* (broom), from the arrangement of the elongate setae on the labellum.

Drosophila wikani **n. sp.**

Fig. 13

DIAGNOSIS. Placed in the *ceratostoma* subgroup based on the lack of a fleshy labellum and transformation of the labellar sclerite into an appendage that is usually directed anteriorly. Distinct from other members of the subgroup by having the labellar sclerite relatively short and unmodified, unbranched, and with few setae.

DESCRIPTION. ♂. **Head.** Front dark brown to black, pale brown near the frontal suture, pollinose; fronto-orbital plates and ocellar triangle slightly darker and

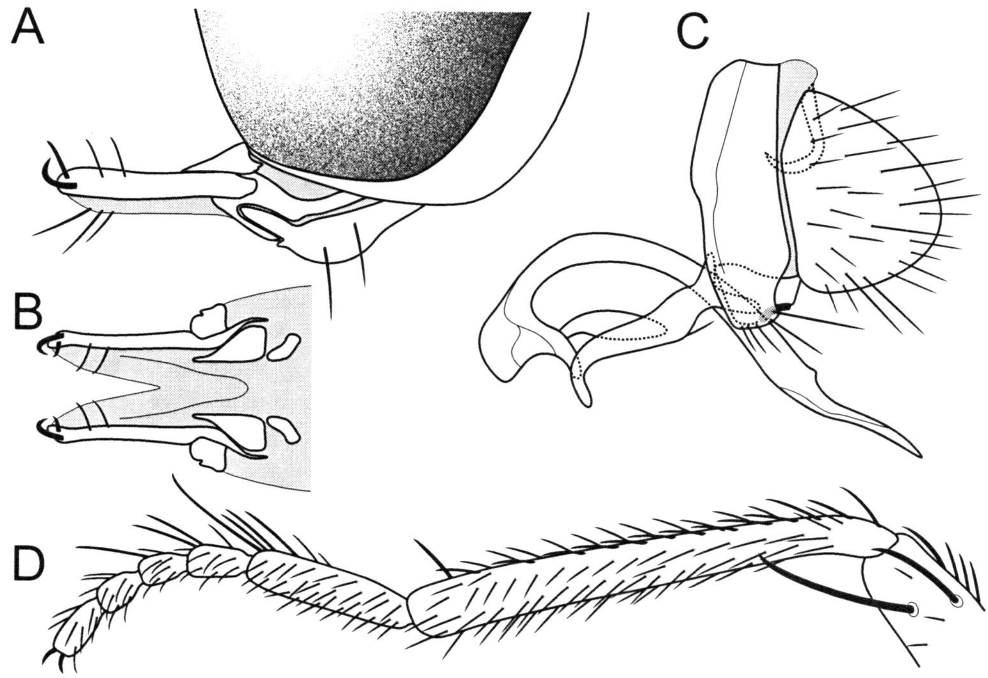

Fig. 13. *Drosophila wikani* male. (A) Lateral view of lower head and labellum in typical position (drawn from pinned specimen); the labellar rim may also be rotated to point posteroventrally toward the mentum, rather than anteriorly as shown. (B) Dorsal view of labellum, in same position as (A). (C) Terminalia. (D) Left front leg, posterior view.

more densely pollinose. Anterior reclinate seta slightly anterior of the proclinate and about twice as long; posterior reclinate stronger but scarcely larger than the anterior. An additional small reclinate seta is present between the proclinate and posterior reclinate setae, about half as long as the former. Ocellar setae about as long as the posterior reclinate; not strongly divergent, often nearly parallel. Vertical setae normal in position, the inner about 3/4 as long as the outer. Eyes with short, inconspicuous pile, about as long as one facet. Face dark brown, with a rounded carina. Antenna brown, paler on the medial surface; arista with 3–4 dorsal and 2 ventral rays in addition to the apical fork. No oral vibrissae. Gena brownish yellow. Palp brown, flattened, only slightly expanded toward the apex, apex broadly rounded, setulose; two strong apical setae. Labellum consisting of only the broad, sclerotized rims, fleshy lobes absent; the rims of each side are connected for about 1/3 their length by a membrane at the dorsal end, and four short, strong, curved setae are present on the ventral third (Fig. 13A, B). The rims are usually held close together and rotated so that the anatomically ventral tip points anterior, giving the appearance of a single sclerotized rod. **Thorax.** Entirely dark brown to black, densely pollinose. Two humeral setae, the ventral about 2/3 as long as the dorsal. Acrostichal setulae in 6 rows. A pair of medial setulae at the transverse suture are enlarged, about twice the length of the other setulae. Two pairs of dorsocentral setae, the anterior about 3/5 as long as the posterior. Halteres pale. **Legs.** Brown, paler on the tibiae and tarsi. Front femur with two long posterior setae near the apex, about 1/3 as long as the femur; setulae of the posterior surface erect. Front tibia lacking elongate cilia, but posterior surface covered with moderately long, erect setulae. Front tarsus covered with erect setulae on the posterior surface; basitarsus with a single long dorsal cilia at the apex and about 5 shorter ones along the apical half; remaining segments each with a pair of cilia at the apex (Fig. 13D). **Wings.** Faintly infuscated anterior of R_{4+5} from the apex to about a level with the dm–cu crossvein, margins diffuse. Costal fringe extending about 1/4–1/3 the distance between the apex of R_{2+3} and R_{4+5}. **Abdomen.** Entirely dark brown, densely pollinose; cerci enlarged, nearly semicircular, about 2/3 as long as high. Epandrium more or less parallel-sided, slightly broadened medially. Aedeagal apodeme rounded trapezoidal, ventral margin concave; about as long as high. Aedeagus with basal angle roughly perpendicular; moderately arched beyond the apodeme, apex narrow and elongate; preapical protuberance present but small (Fig. 13C).

♀. Identical to the male with the following exceptions. **Head.** Posterior reclinate seta about 1.5 times as long as the anterior. Inner vertical seta slightly longer than the outer. One strong oral vibrissa. Palpi with only one apical seta. Labellum unmodified. **Legs.** Front legs without elongate cilia or setae. **Abdomen.** Ovipositor elongate triangular, apex rounded.

MEASUREMENTS. N = 4♂. TL = 1.06 (1.01–1.15) mm; WL = 2.49 (2.28–2.71) mm; TL/WL = 0.4; HW = 0.85 (0.79–0.92) mm; HW/FS = 2.1 (2.1–2.2); HW/TL = 0.8; CI = 3.9 (3.6–4.2); 4V = 1.6 (1.5–1.7); 5X = 1.4 (1.3–1.6); 4C = 0.6; M = 0.4 (0.4–0.5). N = 1♀. TL = 1.22 mm; WL = 2.71 mm; TL/WL = 0.4; HW = 0.90 mm; HW/FS = 2.3; HW/TL = 0.7; CI = 4; 4V = 1.6; 5X = 1.6; 4C = 0.6; M = 0.5.

TYPES. HAWAIʻI: Holotype ♂ (BPBM 16705) and allotype ♀, Humuʻula Saddle, 6650 ft., K47, 13.ix.1967, HLC (♂) & KYK (♀).

PARATYPES. 2♂, Bird Park, Kīlauea, K44, 12.ix.1967, KYK. 3♂, same data as holotype, HLC & KYK. 2♂, Pōhakuloa, 17.vii.1969, KYK. 4♂, Bird Park, N60, 20.ii.1973, SLM. 2♂, Greenwell Ranch, Pauahi, S91, 27.vi.1974, KYK (all UHIM).

DISTRIBUTION & ECOLOGY. Hawaiʻi; found mainly in mesic to dry forest. Breeding hosts unknown; however, based on the collecting sites and potential host plants found there, it is likely to breed on *Myoporum sandwicense* (naio, Myoporaceae).

ETYMOLOGY. From the Hawaiian *wīkani*, hard, rigid, or inflexible, referring to the lack of a fleshy labellum.

DISCUSSION. A single specimen has been collected from Maui (Haleakalā National Park, Kaupō Trail, 5500 ft., on & under *Acacia koa*, 2.viii.2007, KNM), which appears to be identical to the Hawaiʻi specimens. It is being placed under this name but is not designated as a paratype.

freycinetiae subgroup

The mouthpart morphology of this subgroup is somewhat heterogeneous, and its monophyly is not certain. The inclusion of *D. anapuu* and *D. comatifemora* is tentative, as their spines are brown rather than black as in the other species.

1. Labellar spines erect, distinct; almost parallel-sided, usually blunt-tipped (Figs. 16A, 17A) .... 2.
— Labellar spines prostrate, sharp-pointed; often teardrop-shaped and appearing zipper-like or similar to a sclerotized rim (Fig. 15A) .... 6.

2. (1) Only the first two spines directed dorsally, remainder curved ventrally. Maui .... *D. asketostoma* Hardy
— All major spines gently curved dorsally, medially (inclinate), or nearly erect .... 3.

3. (2) Labellum with 6 round, brown spines, one or two distinctly shorter than the others (Fig. 14A). Hawai‘i .... *D. anapuu* n. sp.
— Labellar spines flattened, usually black, all median and ventral spines of similar length .... 4.

4. (3) Labellum with 3 spines, the 2 ventral ones laterally flattened and inclinate (Fig. 16A). Maui .... *D. kualii* n. sp.
— Labellum with 6 or more spines, all flattened dorsoventrally, and slightly curved dorsally (except the dorsal spine which may be inclinate; Fig. 17A) .... 5.

5. (4) Front coxa densely and conspicuously setose over most of the anterior surface. Maui, Moloka‘i .... *D. hirticoxa* Hardy
— Front coxa not densely setose, with about three setae near the apex and several small, inconspicuous setulae scattered over the anterior surface. O‘ahu .... *D. nalomano* n. sp.

6. (1) Labellar spines orange-brown; front femur with a row of elongate antero-ventral setae on the basal two-thirds. Maui .... *D. comatifemora* Hardy
— Labellar spines black; no elongate anteroventral femoral setae .... 7.

7. (6) Front basitarsus normal, without modifications. Kaua‘i .... *D. prominens* Hardy
— Front basitarsus with a short finger-like lobe at the base on the postero-ventral surface (Fig. 15D) .... 8.

8. (7) Dorsum of thorax and abdomen almost entirely brown. O‘ahu .... *D. freycinetiae* Hardy
— Mesonotum yellow (sometimes with a tinge of brown between the dorsocentral setae), abdominal terga narrowly brown medially and yellow laterally. Maui, Lāna‘i? .... *D. dentilabia* n. sp.

Drosophila anapuu **n. sp.**
Fig. 14

DIAGNOSIS. The mouthparts are unique: somewhat resembling *D. nalomano* in having six closely-placed, similar-shaped spines on the labellum, but they are brown and round in cross-section, as in members of the *scolostoma* subgroup.

DESCRIPTION. ♂. **Head.** Front brown, pale anteriorly; fronto-orbital plates and ocellar triangle darker and conspicuously pollinose. Anterior reclinate seta slightly posterior of the proclinate, about half as long; posterior reclinate about 4 times as long as the anterior. Ocellar setae about as long as the posterior reclinate. Vertical setae normal in position, about as long as the posterior reclinate. Eyes with short, inconspicuous pile, about as long as one facet. Face nearly white, with a rounded carina medially. Antenna yellow, tinged with brown; arista with 5–6 dorsal and 2–3 ventral rays in addition to the apical fork. No oral vibrissae, all oral setulae similar. Gena yellow, narrowly brown along the oral margin. Palp yellow, flattened, only slightly expanded toward the apex, apex broadly rounded, setulose; with an elongate apical seta. Labellum with several hair-like setae dorsally; medially with 6 strong, brown, round, pointed spines, slightly curved dorsally, the 5th and sometimes 4th setae about half as long as the others; and ventrally with 2 intermediate setae (Fig. 14A). **Thorax.** Brown, pollinose; katepisternum somewhat paler. Dorsal-most katepisternal setula enlarged, about 2/3 as long as the anterior katepisternal seta. Two humeral setae, the ventral about 2/3 as long as the dorsal. Acrostichal setulae in 6–8 irregular rows. Two pairs of dorsocentral setae, the anterior about 2/3 as long as the posterior. Halteres yellow, stems brownish. **Legs.** Light brown, tibiae and tarsi slightly darker. Front tibia lacking ciliation. Front basitarsus with about 6–7 anterodorsal cilia in two irregular rows along the apical half; second tarsal segment with 1 cilia at the middle and 3 near the apex (Fig. 14C). **Wings.** Hyaline to evenly faintly infuscate. Costal fringe extending 2/5 the distance between the apex of R_{2+3} and R_{4+5}. **Abdomen.** Brown, each segment paler anterolaterally. Epandrium almost parallel-sided, slightly concave along the posterior margin, anteroventrally with a broad spine. Aedeagal apodeme slightly longer than high, trapezoidal. Aedeagus narrow at the base, evenly curved and fishhook-shaped, apex broad and blunt; preapical protuberance moderately large and rounded, near the apex (Fig. 14B).

♀. Unknown.

MEASUREMENTS. N = 2♂. TL = 1.48 (1.46–1.51) mm; WL = 3.16 (3.05–3.28) mm; TL/WL = 0.5; HW = 1.02 (0.99–1.05) mm; HW/FS = 2.2 (2.2–2.3); HW/TL = 0.7; CI = 4.8 (4.8–4.9); 4V = 1.3 (1.3–1.4); 5X = 1.5 (1.4–1.5); 4C = 0.5; M = 0.4.

TYPES. HAWAI'I: Holotype ♂ (BPBM 16706), Upper 'Ōla'a Forest, U13, 19.vii.1976, KYK.

PARATYPES. HAWAI'I: 1♂, Hōnaunau Forest Reserve, 2100 ft., reared ex *Freycinetia* stems, 20.vii.1966, WBH (UHIM).

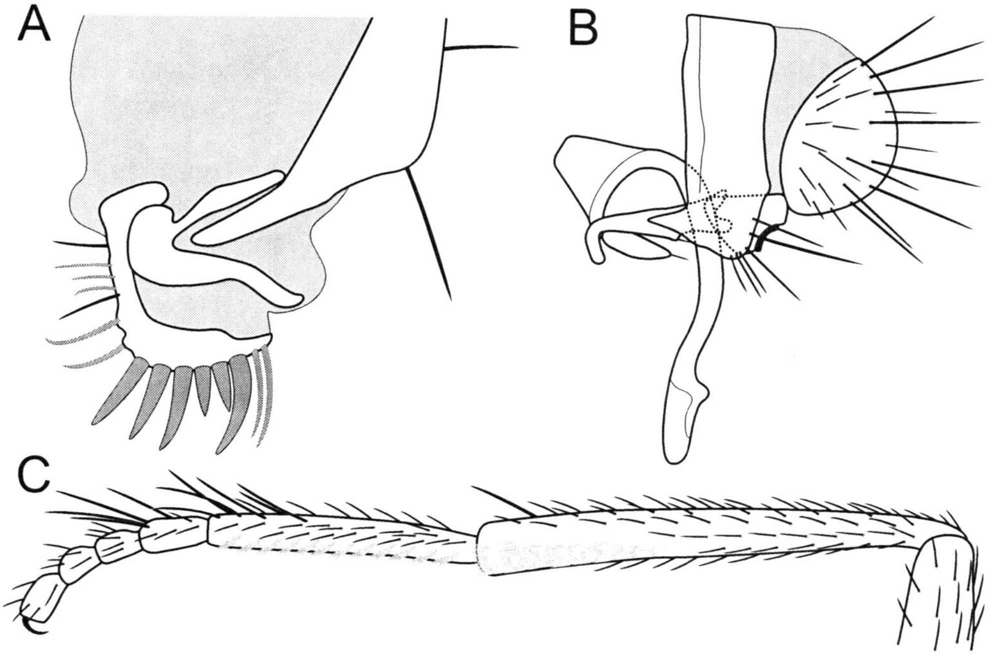

Fig. 14. *Drosophila anapuu* male (holotype from ʻŌlaʻa). (A) Lateral view of labellum. (B) Terminalia. (C) Right front leg, anterior view.

DISTRIBUTION & ECOLOGY. Hawaiʻi; extremely rare, known only from two specimens. Reared from stems of *Freycinetia arborea* (ʻieʻie, Pandanaceae).

ETYMOLOGY. From the Hawaiian *ʻanapuʻu*, uneven, referring to the differing lengths of the labellar spines.

DISCUSSION. This species appears to belong in the *freycinetiae* subgroup based on the gestalt of the mouthparts and the form of the aedeagus, but it could not be placed with it in the key due to the labellar spines being brown rather than black. The two specimens differ in the mouthparts (the holotype has two shortened spines, while the paratype has only one) and may represent actual or incipient sibling species. However, with only two specimens there is no basis for judging variability in the mouthparts (which has been observed in some species, such as *D. aquila*; Magnacca and O'Grady 2007).

Drosophila asketostoma Hardy

Drosophila asketostoma Hardy, 1965:163–165. Holotype ♂, Haleakalā Crater, 8600 ft., viii.1956, on silversword flowers, DEH. BPBM 6301 [examined].

DIAGNOSIS. Mouthparts very distinctive, with only the first strong spine curved dorsally, the second more or less erect and the remaining 4 strongly curved ventrally. Also differs from most other members of the subgroup by having the thorax dark and densely pollinose, appearing gray.

DESCRIPTION. ♂. Front brown above, grading to yellow below the ocellar triangle, pollinose; fronto-orbital plates and ocellar triangle dark, obscured by dense pollen. Anterior reclinate seta slightly posterior of the proclinate and smaller, about half as long; posterior reclinate strong, about 3 times as long as the anterior. Ocellar setae slightly longer than the posterior reclinate. Vertical setae normal in position, the inner about as long as the posterior reclinate, the outer slightly longer. Eyes with short, inconspicuous pile, about as long as one facet. Face pale yellow, with a broad, flat-topped carina medially. Antenna yellow-brown; arista with 3–4 dorsal and 2 ventral rays in addition to the apical fork. Three to four moderately strong oral vibrissae present, not strongly differentiated from the other oral setae which are only slightly shorter. Gena yellow. Palp yellow, flattened, almost parallel-sided, sparsely setulose; one strong apical and one subapical seta, the latter about 2/3 as long as the former. Labellum with 6 dorsoventrally flattened, strong black spines; the first (dorsal-most) strongly curved dorsally, second more or less perpendicular to margin of labellar sclerite, and remaining spines curved ventrally. **Thorax.** Dark, the true ground color obscured by very dense pollen; generally appearing gray, but with posterolateral corners of mesonotum, a stripe along dorsal margin of pleura, and anterior surface of katepisternum appearing reddish brown; humeri yellow. One strong humeral seta; a weak seta that is slightly longer and stronger than surrounding setulae is also present, but the specimens vary in whether this is found dorsally or ventrally of the strong seta. Acrostichal setulae in 6–8 irregular rows. Two pairs of dorsocentral setae, the anterior about 2/3 as long as the posterior. Halteres yellow, stems tinged with brown. **Legs.** Coxae and femora brown, tibiae and tarsi yellow. Front tibia lacking conspicuous cilia, although a few scattered, short, fine, dorsal cilia are present; preapical dorsal seta unusually long, about twice as long as width of tibia at its attachment. Front basitarsus with 10–12 cilia in 2 irregular dorsal rows along the entire length, those at the base very long, rapidly becoming shorter toward the apex; second segment with 3–4 dorsal cilia near the apex; remaining segments with some short erect hairs but without distinct cilia. **Wings.** Mostly hyaline, infuscated over the dm–cu crossvein and in small inconspicuous spots at the apices of R_{2+3}, R_{4+5}, and M. Costal fringe extending 1/4–1/3 the distance between the apex of R_{2+3} and R_{4+5}. **Abdomen.** Dark brown, densely pollinose but the ground color showing through anterolaterally on each segment.

♀. Identical to the male with the following exceptions. **Head.** Front dark brown as far down as halfway between median ocellus and frontal suture, reddish brown below. Face dull yellow. Third segment of antenna dark brown, strongly contrasting to the yellow-brown second segment. Palp brown. Labellum unmodified. **Thorax.** Mesonotum appearing more distinctly reddish brown, with faint gray longitudinal stripes. **Legs.** Tibiae and tarsi tinged with brown. Front legs without cilia. **Abdomen.** Ovipositor more or less parallel-sided but broad, apex blunt.

MATERIAL EXAMINED. MAUI: 2♂ paratypes, same data as holotype (UHIM). 2♀, Haleakalā, 17.vi.1969, on silversword [*Argyroxiphium sandwicense*], W. C. Mitchell (UHIM).

DISTRIBUTION & ECOLOGY. Maui, Haleakalā crater volcanic cinder desert. Appears to be associated with *Argyroxiphium sandwicense*, but has never been reared.

DISCUSSION. The male mouthparts, genitalia, and front leg, and ovipositor were illustrated by Hardy (1965, Fig. 39). Records of the species are the highest-elevation for any Hawaiian *Drosophila* species, and it appears to be restricted to dry, relatively barren habitat – a sharp contrast to virtually all other species, most of which are moisture-limited to mesic and wet forests. A *Scaptomyza* species, *S. (Elmomyza) latitergum*, is also associated with *Argyroxiphium* flowers at Haleakalā (Hardy 1965).

Drosophila n. sp. nr. *asketostoma*?

Three specimens, labelled as "Hawai'i, Pu'u Wa'awa'a, S. L. Montgomery", that appear virtually identical to *D. asketostoma* from Maui are in the UHIM. The apparent association of *D. asketostoma* with Haleakalā Crater and *Argyroxiphium* (although it has never actually been reared; Hardy 1965) makes Pu'u Wa'awa'a a surprising location for it or a sibling species. The lack of a date raises the possibility that the specimens are mislabelled.

Drosophila comatifemora Hardy

Drosophila comatifemora Hardy, 1965:218–219. Holotype ♂ and allotype ♀, Waikamoi, 4000 ft., vii.1956, R. Namba and DEH. BPBM 6329 [examined].

DIAGNOSIS. This species can be distinguished from all others by the combination of a fringe of elongate ventral cilia on the front femur and a row of pale, appressed, flattened spines on the labellum.

DESCRIPTION. ♂. **Head.** Front brown, slightly paler anteriorly; fronto-orbital plates and ocellar triangle yellow, not noticeably pollinose. Anterior reclinate seta slightly posterior of the proclinate and smaller, 1/2–2/3 as long; posterior reclinate strong, about 3 times as long as the anterior. Ocellar setae about as long as the posterior reclinate. Vertical setae normal in position, about as long as the posterior reclinate. Eyes with short, inconspicuous pile, about as long as one facet. Face yellow, broadly convex medially. Antenna yellow, third segment partly brown; arista with 6–7 dorsal and 3 ventral rays in addition to the apical fork. No distinct oral vibrissae present. Gena yellow. Palp yellow, flattened, almost parallel-sided, sparsely setulose; one strong apical seta. Labellum medially with 4 flattened, prostrate, closely-placed, dorsally-directed spines; all closely appressed to the labellar sclerite, difficult to distinguish and appearing similar to a sclerotized rim. Labellum ventrally with several hair-like setae. **Thorax.** Mesonotum yellow, faintly tinged with brown. Pleura yellow, with a brown stripe along dorsal margin of anepisternum and brown spots at dorsal margin of katepisternum and anterior margin of anepimeron. One humeral seta, with no indication of a second. Acrostichal setulae in 6–8 irregular rows. Two pairs of dorsocentral setae, the anterior nearly as long as the posterior. Halteres yellow. **Legs.** Yellow. Front femur with a ventral and anteroventral row of elongate cilia on the basal half (the rows with about 6 and 8 hairs respectively), the longest about 1.5–2 times as long as greatest width of femur, becoming shorter apically; two long posterior setae near the apex, the dorsal one about 1/3 as long as the femur, ventral about half as long as the dorsal; setulae of the posterior surface erect. Front tibia lacking elongate cilia (three short, fine, inconspicuous cilia present: one near the base, one near the midpoint, and one at 2/3 the length), but posterior surface covered with moderately long, erect setulae. Front tarsus covered with erect setulae on the posterior surface; basitarsus with about 12–15 elongate cilia in three irregular dorsal to anterodorsal rows along the entire length; second segment with 5 cilia, remaining segments each with three. **Wings.** Hyaline; faintly infuscated generally, but without markings except sometimes a small mark visible over the dm–cu crossvein. Costal fringe extending about 2/5 the distance between the apex of R_{2+3} and R_{4+5}. **Abdomen.** Distinctly banded, each segment brown medially and along the posterior margin, broadly yellow anterolaterally. Epandrium parallel-sided, elongate, nearly as long as the cerci.

♀. Identical to the male with the following exceptions. **Head.** Labellum unmodified. **Legs.** Front legs without cilia. **Abdomen.** Ovipositor broadly triangular, apex blunt.

MATERIAL EXAMINED. MAUI: 1♂, Kīpahulu Valley, 3000 ft., L1, 3.viii.1967, HLC. 1♂, Kīpahulu Valley, 2500 ft., L3, 4.viii.1967, HLC. 1♂, ridge above Kīpahulu Valley, J88, 21.viii.1967, KYK (all above at UHIM). 2♂ 4♀, Ha'ikū Uka, Heed Trail, 4200 ft., sweeping vegetation and ground, 31.vii.2007, KNM (UCPO). 2♂ 1♀, Kaupō Trail, 5000 ft., sweeping vegetation and ground, 4.viii.2007, KNM (UCPO). 1♂, Ha'ikū Uka, Heed Trail, 4200 ft., sweeping vegetation and ground, 6.viii.2007, KNM (UCPO).

DISTRIBUTION & ECOLOGY. Maui, wet and mesic forest. Breeding habits unknown.

DISCUSSION. The male front leg, wing, and genitalia were illustrated by Hardy (1965, Fig. 65); the mouthparts were illustrated by Magnacca and O'Grady (2007, Fig. 2b).

Drosophila dentilabia n. sp.

Fig. 15

DIAGNOSIS. Very similar to *D. freycinetiae*, with a basal lobe on the basitarsus, but with the thorax all yellow (sometimes tinged with brown between the dorsocentral setae) instead of brown.

DESCRIPTION. ♂. **Head.** Front pale brown; fronto-orbital plates and ocellar triangle yellow. Anterior reclinate seta slightly posterior of the proclinate and smaller, 2/3–3/4 as long; posterior reclinate strong, about 2 times as long as the anterior. Ocellar setae about as long as the posterior reclinate. Vertical setae normal in position, about as long as the posterior reclinate. Eyes with short, inconspicuous pile, about as long as one facet. Face yellow, nearly flat, short, membranous below level of antennal apex. Antenna light brown; arista with 4–6 dorsal and 2 ventral rays in addition to the apical fork. Three to four moderately strong oral vibrissae present, not strongly differentiated from the other oral setae which are only slightly shorter. Gena yellow, narrowly brown along the oral margin. Palp yellow, flattened, slightly clavate, broadest near the apex, rather densely setulose over the apical 2/3; one strong apical and one subapical seta. Labellum with 9 flattened, appressed, teardrop-shaped, black spines (Fig. 15A). **Thorax.** Mesonotum brownish yellow; often with a narrow brown median vitta over the posterior half. Pleura largely yellow with a brown stripe along the dorsal margin of the anepisternum and tinged with brown over the anepimeron. Two humeral setae, the ventral about 2/3 as long as the dorsal. Acrostichal setulae in 6–8 irregular rows. Two pairs of dorsocentral setae, the anterior about 2/3 as long as the posterior. Halteres yellow. **Legs.** Yellow. Front tibia lacking conspicuous ciliation but with a few (4–6) short, thin, scattered dorsal and posterodorsal cilia, only about as long as setulae. Front basitarsus with 12–15 cilia in 3 irregular dorsal rows along the apical 3/4; second segment with 2 dorsal cilia near the apex; remaining segments without distinct cilia (Fig. 15C). Front basitarsus with a posterior lobe at the base, about 1/3 as long as the basitarsus; apex of lobe and facing surface of basitarsus covered with short black spines (Fig. 15D). **Wings.** Mostly hyaline, with diffuse infuscation anteroapically along R_{2+3}, R_{4+5}, and M, and over the dm–cu crossvein. Costal fringe extending half the distance between the apex of R_{2+3} and R_{4+5}. **Abdomen.** Yellow laterally, brown medially (T1 entirely yellow), the brown area narrowing posteriorly; cerci and extreme lateral margins of tergites (on ventral surface) also brown. Epandrium parallel-sided, elongate, about as long as the cerci. Aedeagal apodeme triangular,

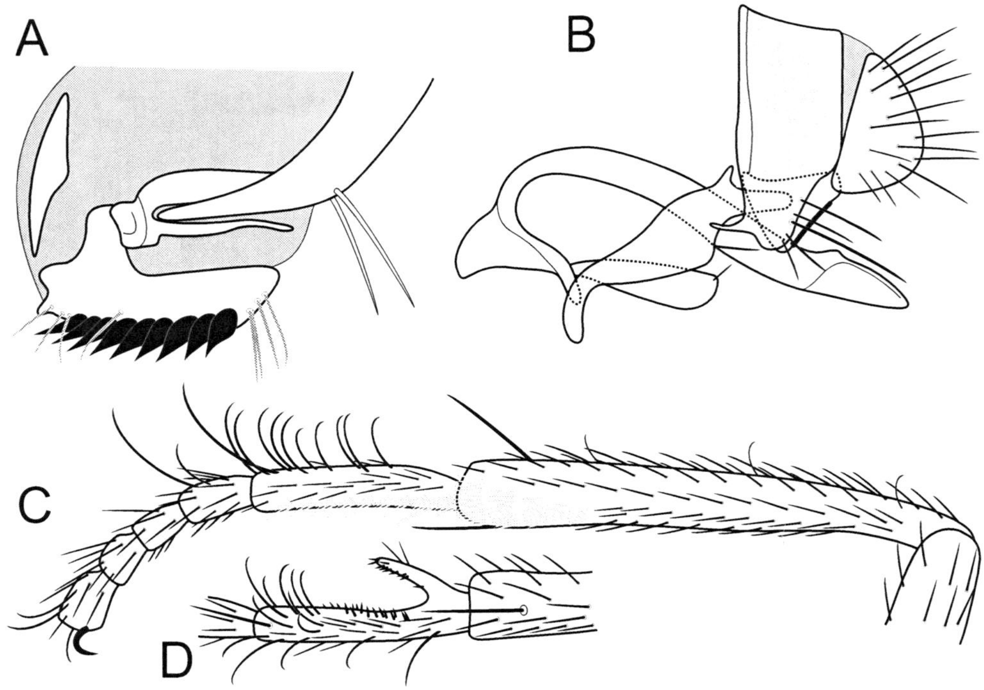

Fig. 15. *Drosophila dentilabia* male. (A) Lateral view of labellum. (B) Terminalia. (C) Right front leg, anterior view. (D) Right basitarsus, dorsal view.

distinctly longer than high. Aedeagus strongly bent at the base, basal angle acute, apex short; preapical protuberance small (Fig. 15B).

♀. Identical to the male with the following exceptions. **Head.** Face not shortened, extending below vibrissal angle. Palp without prominent subapical seta. Labellum unmodified. **Thorax.** Mesonotum broadly tinged with brown between dorsocentral setae. **Legs.** Front legs without elongate cilia or setae, basitarsus unmodified. **Abdomen.** Most of dorsal surface brown, yellow only on the sides. Ovipositor yellow, broadly triangular, apex blunt.

MEASUREMENTS. N = 3♂. TL = 1.28 (1.20–1.33) mm; WL = 2.49 (2.38–2.60) mm; TL/WL = 0.5; HW = 1.02 (0.96–1.05) mm; HW/FS = 2.4 (2.3–2.4); HW/TL = 0.8; CI = 4.4 (4.2–4.6); 4V = 1.3 (1.2–1.4); 5X = 1.1 (1.0–1.3); 4C = 0.5; M = 0.3 (0.3–0.4). N = 1♀. TL = 1.20 mm; WL = 2.53 mm; TL/WL = 0.5; HW = 0.98 mm; HW/FS = 2.4; HW/TL = 0.8; CI = 4.1; 4V = 1.5; 5X = 1.3; 4C = 0.6; M = 0.4.

TYPES. MAUI: Holotype ♂ (BPBM 16694) and allotype ♀, Pu‘u Kukui Trail, 2900–3700 ft., sweeping *Freycinetia*, 8.viii.2007, KNM.

PARATYPES. 2♂, same data as holotype (UHIM).

DISTRIBUTION & ECOLOGY. Maui and probably Lāna‘i. Apparently associated with *Freycinetia*.

ETYMOLOGY. From the Latin *denta* (tooth) + *labia* (lip), referring to the closely-placed spines of the labellum.

DISCUSSION. A specimen from Lāna'i (11.iii.1973, 2400 ft., "*Freycinetia* axil with fruit and rat shit," SLM) with a basal lobe on the front tarsus is probably this species, but it was not available for comparison.

Drosophila freycinetiae Hardy

Drosophila freycinetiae Hardy, 1965:277–279. Holotype ♂ and allotype ♀, Mt. Tantalus, on *Freycinetia*, viii.1952, DEH and M. S. Adachi. BPBM 6360 [examined].

DIAGNOSIS. Distinguished by most other species by the basal lobe of the basitarsus; separated from *D. dentilabia* by the darker coloration of the thorax and abdomen.

DESCRIPTION. ♂. **Head.** Front brown, grading to yellow anteriorly; fronto-orbital plates and ocellar triangle darker brown. Anterior reclinate seta about even with the proclinate and smaller, about 3/5 as long; posterior reclinate strong, about twice as long as the anterior. Ocellar setae about as long as the posterior reclinate. Vertical setae normal in position, about as long as the posterior reclinate. Eyes with short, inconspicuous pile, about as long as one facet. Face yellow to white, short, ventral third membranous; with a low, rounded median carina. Antenna yellowish brown; arista with 4–5 dorsal and 2 ventral rays in addition to the apical fork. Three to four moderately strong oral vibrissae present, not strongly differentiated from the other oral setae which are only slightly shorter. Gena yellow, narrowly brown along the oral margin. Palp yellow, flattened, slightly clavate, broadest near the apex, moderately setulose over the apical 2/3; one strong apical and one subapical seta. Labellum with 9 flattened, appressed, teardrop-shaped, black spines. **Thorax.** Mesonotum brown, humeri yellow ventrally. Pleura largely yellow with a brown stripe along the dorsal margin of the anepisternum and tinged with brown over the anepimeron. Two humeral setae, the ventral 1/2–2/3 as long as the dorsal. Acrostichal setulae in 6 regular rows. Two pairs of dorsocentral setae, the anterior about 2/3 as long as the posterior. Halteres yellow. **Legs.** Yellow. Front tibia lacking conspicuous ciliation but with a few short, thin, scattered dorsal and posterodorsal cilia, only about as long as setulae. Front basitarsus with 8–10 cilia in two irregular dorsal rows along the apical 3/4; remaining segments with 2–3 dorsal cilia near their apices, those of segments 3–5 sometimes not very distinct from other setulae. Front basitarsus with a posterior lobe at the base, about 1/3 as long as the basitarsus; apex of lobe and facing surface of basitarsus covered with short black spines. **Wings.** Mostly hyaline, with diffuse infuscation anteroapically along R_{2+3}, R_{4+5}, and M, and over the dm–cu crossvein. Costal fringe extending half the distance between the apex of R_{2+3} and R_{4+5}. **Abdomen.** Mostly brown, posterior tergites yellow laterally; cerci and extreme lateral margins of tergites (on ventral surface) also brown. Epandrium parallel-sided, elongate, about as long as the cerci. ♀. Identical to the male with the following exceptions. **Head.** Face not shortened, extending below vibrissal angle. Palp without prominent subapical seta. Labellum unmodified. **Thorax.** Mesonotum broadly tinged with brown between dorsocentral setae. **Legs.** Front legs without elongate cilia or setae, basitarsus unmodified. **Abdomen.** Almost entirely brown, small lateral areas tinged with yellow. Ovipositor brown, broadly triangular, apex blunt.

MATERIAL EXAMINED. O'AHU: 1♂ paratype, Mānoa Cliff Trail, 25.vii.1917 [no collector]. 1♂ paratype, Mt. Tantalus, v.1953, M. S. Adachi. 1♂ paratype, Mt. Tantalus, vi.1953, DEH. 1♂ 1♀ paratypes, Mt. Tantalus, "*Freycinetia*," viii.1955, DEH. 1♂, Mānoa Cliff Trail, ex *Freycinetia* single leaf base, 10.i.1970, WBH. 1♂, Wiliwilinui Ridge, "(P34? 27.vi.1970)," SLM (all UHIM).

DISTRIBUTION & ECOLOGY. O'ahu. Associated with rosettes of *Freycinetia*, the larvae apparently feeding on detritus that accumulates among the leaf bases.

DISCUSSION. The male mouthparts, front leg, wing, and genitalia were illustrated by Hardy (1965, Fig. 96).

Drosophila hirticoxa Hardy

Drosophila hirticoxa Hardy, 1965:302–303. Holotype ♂, Maui, Haleakalā Crater, Palikū, 6500 ft., vi.1953, C. R. Joyce. BPBM 6370 [examined].

DIAGNOSIS. Labellum with elongate, erect spines, similar to *D. nalomano* but longer. Easily distinguished from that species by the densely setose anterior surface of the front coxa.

DESCRIPTION. ♂. **Head.** Front brown above, yellow anteriorly; fronto-orbital plates and ocellar triangle dark brown, conspicuously pollinose. Anterior reclinate seta slightly posterior of the proclinate and small, about 2/5 as long or less; posterior reclinate strong, about 3–4 times as long as the anterior. Ocellar setae about as long as the posterior reclinate. Vertical setae normal in position, about as long as the posterior reclinate. Eyes with short, inconspicuous pile, about as long as one facet. Face yellowish brown, with a low, rounded median carina. Antenna yellow; arista with 4 dorsal and 2 ventral rays in addition to the apical fork. Six strong oral vibrissae present. Gena yellow, narrowly brown along the oral margin. Palp yellow, flattened, slightly clavate, broadest at about 2/3 the length, apical half densely setulose; one strong apical seta. Labellum with about 11 dorsoventrally flattened, erect, black spines, about equal in length; all gently curved dorsally, nearly straight. **Thorax.** Brown, densely gray pollinose; humerus and propleura yellow. Two humeral setae, the ventral about 3/4 as long as the dorsal. Acrostichal setulae in 6–8 irregular rows; a pair of elongate acrostichal setae present near the midline just anterior of the suture. Two pairs of dorsocentral setae, the anterior about 2/3 as long as the posterior. Halteres brown. **Legs.** Yellow. Anterior surface of front coxa densely covered with black setulae. Front tibia lacking ciliation. Front tarsus with elongate dorsal cilia over the first three segments; remaining segments with shorter hairs. Posterior surfaces of front tibia and tarsus covered with short, thin, erect setulae. **Wings.** Entirely hyaline. Costal fringe extending 1/5 the distance between the apex of R_{2+3} and R_{4+5}. **Abdomen.** Brown, pollinose. ♀. Unknown. A female collected at the same locality as the holotype a year earlier was described by Hardy (1965) but not included as an allotype, and its association is uncertain.

MATERIAL EXAMINED. MAUI: 1♂, Palikū, Haleakalā Crater, 6500 ft., vii.1963, LHT. 1♂, Palikū, 6.vii.1964, MPT81 (no collector). 1♂, Palikū, Haleakalā Crater, 27.vii.1964, D. Gubler. 1♂, Palikū, Haleakalā Crater, 12.vii.1966, KYK. MOLOKA'I: 1♂, Pu'u Kolekole, 3600 ft., 20.vii.1964, HLC (all UHIM).

DISTRIBUTION & ECOLOGY. Maui and Moloka'i; rare. Breeding habits unknown.

DISCUSSION. The male head, front coxa and tarsus, and genitalia were illustrated by Hardy (1965, Fig. 110). The abdomen was originally described as "Almost entirely yellow, tinged with brown on terga two to four," but the UHIM specimens appear to have the color much better preserved than the holotype and have the abdomen dark, similar to the thorax.

Drosophila kualii n. sp.

Fig. 16

DIAGNOSIS. Mouthparts somewhat similar to *D. asketostoma*, but with only three black spines, two of which are directed medially rather than dorsally or ventrally. Body also much smaller.

DESCRIPTION. ♂. **Head.** Front brown on upper 2/3, yellow below the level of the ocellar triangle. Anterior reclinate seta slightly posterior of the proclinate, about 3/5 as long; posterior reclinate about 3 times as long as the anterior. Ocellar setae about as long as the posterior reclinate. Vertical setae normal in position, about as long as the posterior reclinate. Eyes with short, inconspicuous pile, about as long as one facet. Face yellow, with a rounded carina medially. Antenna brown on third segment, paler basally; arista with 3 dorsal and 1 ventral ray (possibly some broken off). One strong oral vibrissa, the next seta 1/2–2/3 as long, remaining oral setulae smaller. Gena yellow, narrowly brown along the oral margin. Palp yellow, flattened, convex on the lateral margin and nearly straight on the medial, broadest at about 2/3 the length, sparsely setulose; two strong apical setae. Labellum with three broad, flattened, erect, black spines, which end in a blunt point (Fig. 16A). **Thorax.** Entirely brown, pollinose. Two humeral setae, the ventral only about half as long as the dorsal. Acrostichal setulae in 6 rows. Two pairs of dorsocentral setae, the anterior about 2/3 as long as the posterior. Halteres yellow. **Legs.** Brown, darkest

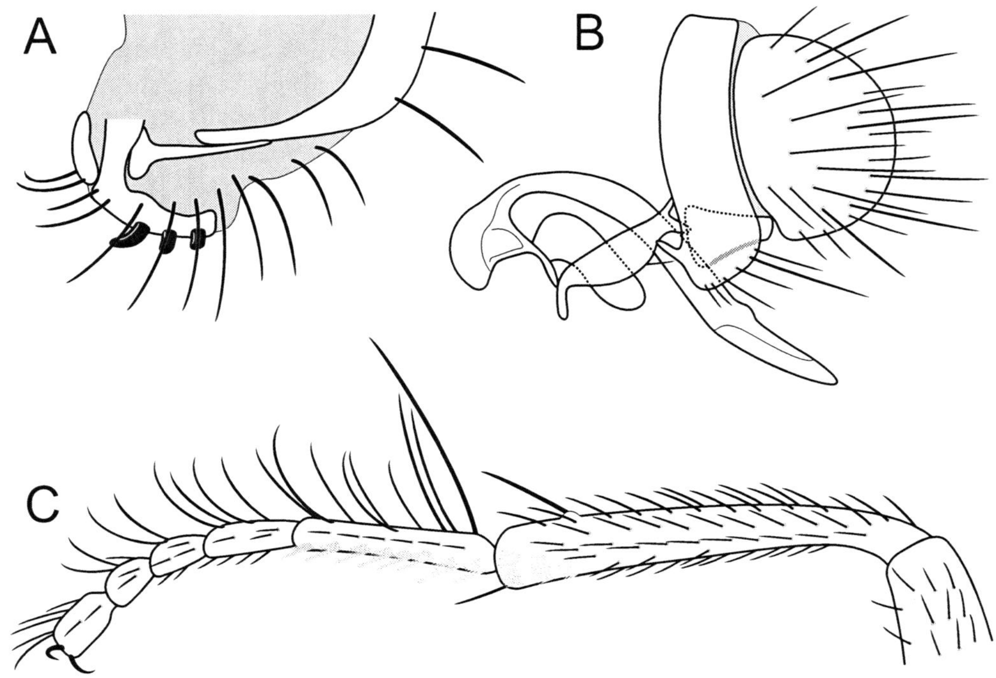

Fig. 16. *Drosophila kualii* male. (A) Lateral view of labellum (drawn from pinned specimen). (B) Terminalia. (C) Right front leg, anteroventral view.

on the femora, yellowish at joints. Front tibia lacking ciliation. Front tarsus with a dorsal and an anterodorsal row of cilia, each with 4 on the basitarsus, 2 on the second segment, and one each on the remaining segments, becoming shorter apically; the first two cilia of the anterodorsal row are much longer than the rest, about as long as the basitarsus (Fig. 16C). **Wings.** Entirely hyaline. Costal fringe extending 1/4 the distance between the apex of R_{2+3} and R_{4+5}. **Abdomen.** Entirely brown. Epandrium more or less parallel-sided, anterior margin slightly sinuate. Aedeagal apodeme rounded quadrate, ventral margin concave; about as long as high. Aedeagus strongly bent at the base, basal angle acute; apex elongate, preapical protuberance present but small (Fig. 16B).

♀. Unknown.

Measurements. N = 2♂. TL = 1.02 (1.00–1.05) mm; WL = 2.29 (2.07–2.50) mm; TL/WL = 0.5 (0.4–0.5); HW = 0.80 (0.75–0.85) mm; HW/FS = 2.3; HW/TL = 0.8; CI = 4.9 (4.1–5.8); 4V = 1.4 (1.2–1.6); 5X = 1.6; 4C = 0.5 (0.4–0.6); M = 0.5.

Types. Maui: Holotype ♂ (BPBM 16707), Haleakalā National Park, Kaupō Trail, 5500 ft., on & under *Acacia koa*, 2.viii.2007, KNM.

Paratypes. 1♂, Palikū, Haleakalā Crater, 6500 ft., CH1.32, 23.vii.1963, WBH (UHIM).

DISTRIBUTION & ECOLOGY. Maui; known only from two specimens. Breeding habits unknown.

ETYMOLOGY. From the Hawaiian *kuali'i*, dwarfed plant or animal, referring to the small size of this species compared to others in the subgroup.

DISCUSSION. This species may be related to *D. asketostoma*, which seems to be associated with silverswords (*Argyroxiphium*, Asteraceae; Hardy 1965). Its exact habitat is uncertain; while many wet forest species are found at Palikū, the presence of *D. asketostoma* in drier areas of Haleakalā Crater suggests that *D. kualii* could be found in the latter as well. The Kaupō site consisted of mesic forest, but few typical hosts were present except *Acacia koa*.

Drosophila nalomano n. sp.

Fig. 17

DIAGNOSIS. Readily distinguished from all other species by the dorsoventrally flattened, erect, black spines on the labellum; these are somewhat similar to those of *D. hirticoxa*, but fewer in number, and lacking the setulose front coxa of that species.

DESCRIPTION. ♂. **Head.** Front brown, paler anteriorly; fronto-orbital plates and ocellar triangle conspicuously pollinose. Anterior reclinate seta slightly posterior of the proclinate and smaller, 1/2–2/3 as long; posterior reclinate strong, about 3 times as long as the anterior. Ocellar setae about as long as the posterior reclinate. Vertical setae normal in position, about as long as the posterior reclinate. Eyes with short, inconspicuous pile, about as long as one facet. Face brown, nearly flat. Antenna brown; arista with 3 dorsal and 2 ventral rays in addition to the apical fork. Two strong oral vibrissae present; 2–3 additional setae near the vibrissal angle are about twice as long as the other oral setae, but not much stronger. Gena yellow, narrowly brown along the oral margin. Palp yellow, flattened, slightly clavate, broadest at about 2/3 the length, sparsely setulose; one strong apical seta. Labellum with 7 flattened, erect, black spines; the first is strongly bent dorsally, the remainder only gently curved (Fig. 17A). **Thorax.** Brown, pollinose; tinged with yellow in the middle of the anepisternum and lower half of the katepisternum. Dorsal-most katepisternal setula enlarged, nearly as long as the anterior katepisternal seta. Two humeral setae, the ventral about 3/4 as long as the dorsal. Acrostichal setulae in 6–8 irregular rows. Two pairs of dorsocentral setae, the anterior about 2/3 as long as the posterior. Halteres brown. **Legs.** Yellow. Front tibia lacking ciliation. Front basitarsus with 6–8 anterodorsal cilia in close pairs along the apical 2/3, and a single posterodorsal cilia at the apex; second segment with 3 dorsal and 1 posterodorsal near the apex; remaining segments with 1–2 short cilia (Fig. 17C). **Wings.** Entirely hyaline. Costal fringe extending 1/3 the distance between the apex of R_{2+3} and R_{4+5}. **Abdomen.** Entirely dark brown. Epandrium trapezoidal in

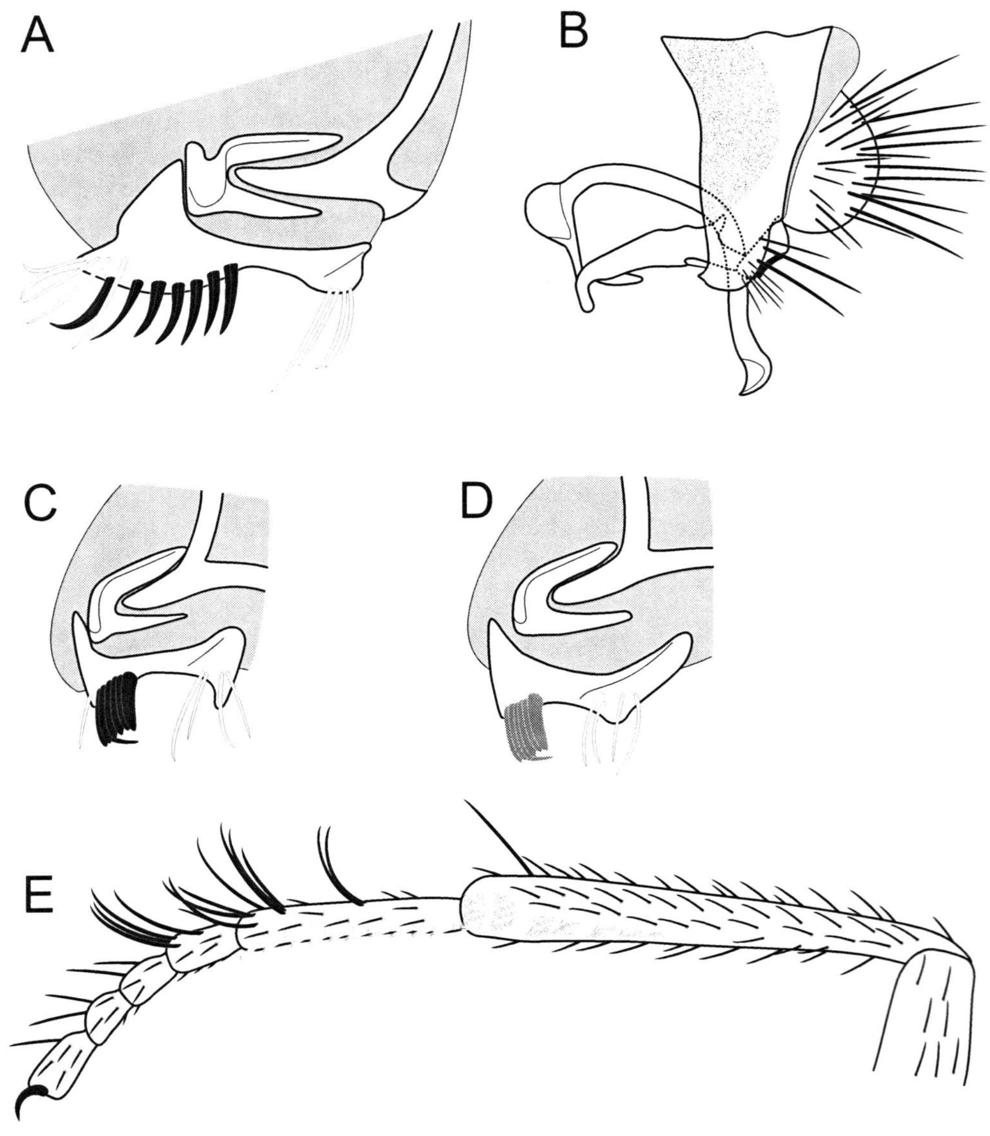

Fig. 17. *Drosophila nalomano* male. (A) Lateral view of labellum. (B) Terminalia. Ventrolateral view of labellum of (C) Kaau Crater specimen and (D) Castle Trail specimen. (E) Right front leg, anterior view.

lateral view, dorsal margin nearly three times as long on the ventral apex; with a small, broad ventroapical spine. Aedeagal apodeme semicircular, nearly as long as high. Aedeagus weakly bent at the base, basal angle obtuse, apex short and strongly arched; preapical protuberance absent (Fig. 17B).
♀. Identical to the male with the following exceptions. **Head.** Arista with only 1 ventral ray. Labellum unmodified. **Thorax.** Dorsal katepisternal setula short, only about half as long as the anterior. Halteres yellow, tinged with brown on the stems. **Legs.** Front legs without elongate cilia or setae. **Abdomen.** Dark brown, each segment paler anterolaterally. Ovipositor yellow, broadly triangular, apex blunt.
MEASUREMENTS. N = 1♂. TL = 1.00 mm; WL = 2.10 mm; TL/WL = 0.5; HW = 0.78 mm; HW/FS = 2.2; HW/TL = 0.8; CI = 3.9; 4V = 1.4; 5X = 1.3; 4C = 0.5; M = 0.4. N = 1♀. TL = 1.22 mm; WL = 2.42 mm; TL/WL = 0.5; HW = 0.87 mm; HW/FS = 2.3; HW/TL = 0.7; CI = 3.8; 4V = 1.6; 5X = 1.2; 4C = 0.6; M = 0.5.

TYPES. O'AHU: Holotype ♂ (BPBM 16708) and allotype ♀, Ka'au Crater rim, 1600 ft., ex rotting *Wikstroemia*, 29.vi.1987, ATO & WDP.

PARATYPES. 1♂, same data as holotype (UHIM).

OTHER MATERIAL. O'AHU: 1♂, Castle Trail, ex rotten stem *Touchardia*, 11.iv.1970 [no collector] (UHIM).

DISTRIBUTION & ECOLOGY. O'ahu; very rarely collected, known only from the Ko'olau range. Reared from bark of *Wikstroemia* ('ākia, Thymeliaceae) and *Touchardia* (olonā, Urticaceae).

ETYMOLOGY. From the Hawaiian *nalo* (fly) + *manō* (shark), referring to the form of the labellar spines.

DISCUSSION. This species has a peculiar apex of the aedeagus; rather than being elongate and straight, it is short and arched. This type is found in *D. freycinetiae* but not *D. prominens*. A similar form has apparently evolved independently in a *picture-wing* species, *D. hirtipalpus*. The Castle Trail specimen is nearly identical except for the ventral portion of the labellum (compare Figs. 17C and D); the spines are also brown, but this may be due to fading of the specimen. Without additional specimens it is unclear whether this is intraspecific variation, an ecological variant, or a distinct species. It is not designated as a paratype.

Drosophila prominens Hardy

Drosophila prominens Hardy, 1965:432–434. Holotype ♂ and allotype ♀, Kaua'i, Kawaikōī Stream, 3700 ft., viii.1953, DEH. BPBM 6428 [examined].

DIAGNOSIS. Labellum with appressed black spines; differing from *D. dentilabia* and *D. freycinetiae* by having only 6 spines, lacking a basal lobe on the basitarsus, and having the pleura entirely brown.

DESCRIPTION. ♂. **Head.** Front dark and pollinose, grading from velvety black near the vertex to reddish brown at the frontal suture; fronto-orbital plates and ocellar triangle darker brown, more densely pollinose. Anterior reclinate seta about even with the proclinate and smaller, 2/5–1/2 as long; posterior reclinate strong, about 3 times as long as the anterior. Ocellar setae about as long as the posterior reclinate. Vertical setae normal in position, about as long as the posterior reclinate. Eyes with short, inconspicuous pile, about as long as one facet. Face dark brown, with a prominent, rounded median carina. Antenna dark brown; arista with 5–6 dorsal and 2–3 ventral rays in addition to the apical fork. Two strong oral vibrissae present, distinct from the other oral setae. Gena yellow, narrowly brown along the oral margin. Palp yellow, flattened, almost parallel-sided, moderately setulose over the apical 2/3; one strong, elongate apical seta, curved near the apex, as long or longer than the palp. Labellum with 6 flattened, appressed, teardrop-shaped, black spines, the dorsalmost with the tip elongate and curved. **Thorax.** Almost entirely dark brown, tinged with yellow on the humeri; moderately pollinose. Two humeral setae, about equal in length. Acrostichal setulae in 6–8 irregular rows. Two pairs of dorsocentral setae, the anterior about 2/3 as long as the posterior. Halteres yellow, stem tinged with brown. **Legs.** Coxae and femora brown, tibiae and tarsi yellowish brown. Front tibia without cilia; preapical dorsal seta unusually long, about twice as long as width of tibia at its attachment. Front basitarsus with 6–8 cilia in 2 irregular anterodorsal rows along the entire length, those at the base very long, rapidly becoming shorter toward the apex; short cilia continued on remaining segments, not always distinct from other setulae. **Wings.** Hyaline; faintly infuscated generally, but without markings. Costal fringe extending 1/4–1/3 the distance between the apex of R_{2+3} and R_{4+5}. **Abdomen.** Mostly brown, each segment tinged with yellow anterolaterally. Epandrium with a broad lobe laterally along the anterior margin. Aedeagal apodeme triangular, distinctly longer than high. Aedeagus strongly bent at the base, basal angle acute, apex elongate; preapical protuberance small.
♀. Identical to the male with the following exceptions. **Head.** Only one strong oral vibrissa present. Apical seta of palp nearly straight, not unusually elongate. Labellum unmodified. **Thorax.** Humeri brown. Katepisternum yellowish brown. **Legs.** Entirely yellow. Front legs without cilia. **Abdomen.** Ovipositor broadly triangular, apex blunt.

MATERIAL EXAMINED. KAUA‘I: 1♂ 1♀ paratype, Halemanu Swamp, viii.1953, DEH. 1 ♀ paratype, Mt. Wai‘ale‘ale Trail, viii.1953, DEH. 1♂, Miloli‘i Valley, 2000 ft., ex *Touchardia*, SLM (all UHIM).

DISTRIBUTION & ECOLOGY. Kaua‘i. Reared from *Touchardia* (olonā, Urticaceae).

DISCUSSION. The male mouthparts, antenna, front tarsus, wing, and genitalia, and ovipositor were illustrated by Hardy (1965, Fig. 173). The color differences between males and females may be real or due to changes in older specimens but may also indicate that they are incorrectly associated; the sexes have never been reared together, and many species occur in the Kōke‘e area.

fuscoamoeba subgroup

All the species of this subgroup are very rarely collected or reared, and it is highly likely that additional species exist. Several species have been reared from rotting fern rachises (Magnacca et al. 2008), which have not been widely examined for *Drosophila*, and new species may be found if more rearing is done from ferns.

1. Mesonotum unicolorous; wing markings relatively faint .... 2.
— Mesonotum vittate; wing markings variable, often with distinct patterns.... .... 4.

2. (1) Front tarsus with elongate cilia, over twice as long as the width of the tarsus. Hawai'i .... *D. aquila* Hardy
— Front tarsus with short cilia, not much longer than the width of the tarsus.. .... 3.

3. (2) Third costal section (between R_1 and R_{2+3}) about twice as long as the fourth. Maui .... *D. furva* Hardy
— Third costal section about 3 times as long as the fourth. Maui, Moloka'i .. .. *D. araiotrichia* Hardy

4. (1) Third costal section equal to or shorter than the fourth, vein R_{2+3} ending almost opposite the dm–cu crossvein; veins R_{4+5} and M not or weakly sinuate .... 5.
— Third costal section 1.5 times or more longer than the fourth, vein R_{2+3} ending beyond the dm–cu crossvein; veins R_{4+5} and M strongly sinuate .... 6.

5. (4) Third costal section about equal to the fourth; anterior margin with three large infuscations: at the base, at the apex of vein R_{2+3}, and at the apex of R_{4+5}. Kaua'i .... *D. agitona* Hardy
— Third costal section shorter than the fourth; infuscations at the apices of vein R_{2+3} and R_{4+5} each interrupted by clear areas at the wing margin; cell c hyaline. O'ahu .... *D. fuscoamoeba* Bryan

6. (4) Pleura entirely brown; front tarsus with elongate cilia, over twice as long as the width of the tarsus. Maui, Moloka'i .... *D. clydonia* Hardy
— Katepisternum largely yellow; front tarsus with short cilia, not much longer than the width of the tarsus, or none. Hawai'i .... 7.

7. (6) Spot on r–m crossvein extended basally in cell br almost to the humeral crossvein; front tarsus with short cilia .... *D. brevicilia* Hardy
— Spot on r–m crossvein small, not extended basally; front tarsus lacking cilia .... *D. megasticta* Hardy

***mitchelli* subgroup** (key modified from Hardy and Kaneshiro 1975b)

Members of this subgroup are among the few *modified mouthparts* species that are commonly attracted to standard banana and mushroom baits; paradoxically, it is also the only subgroup for which there are no rearing records (Magnacca et al. 2008).

1. Wing infuscated anteroapically. Hawai'i ....2.
— Wing hyaline or evenly infuscated, without distinct markings..................3.

2. (1) Labellum with five black spines arranged in two groups; femora brown..... ....*D. mitchelli* Hardy
— Labellum with three black spines in one group; femora yellow.................. ....*D. nigrocirrus* Hardy

3. (1) Three dorsal spines of the labellum separated from the ventral-most spine by a gap, spines not bifid. Maui ....*D. hystricosa* Hardy and Kaneshiro
— Labellar spines placed close together in one group, at least some spines bifid on close inspection ....4.

4. (3) Femora yellow; labellum truncate below spines. O'ahu ....*D. biseriata* Hardy
— Femora dark; labellum ventrally produced beyond spines. Kaua'i ....*D. furvifacies* Hardy

***nanella* subgroup**

Three species formerly placed in the *bridwelli* subgroup, *D. albifacies*, *D. curticilia*, and *D. dolomata*, are transferred to this subgroup. They are united by having a single strong, relatively straight spine arising near the middle of the labellar sclerite, followed by several thinner spines that grade into hair-like setae, as opposed to a cluster of medially-curved spines at the dorsal end in the *semifuscata*/*bridwelli* subgroup; the epandrium with a prominent rounded anterior lobe laterally, which sometimes also bears a pointed secondary projection dorsally; and the aedeagus with the basal apodeme longer than high and the basal angle acute (Fig. 18B). At least two of the species, *D. dolomata* and *D. nanella*, are associated with *Pisonia* leaves (Magnacca et al. 2008).

1. Front basitarsus very short, shorter than the next three segments combined; distitarsus usually held tightly curled, each segment with a brush of short, stout setulae ventrally. Kaua'i..*D. nanella* Hardy
— Front basitarsus equal to at least the next three segments; distitarsus not held curled up, lacking brush-like ventral setulae ....2.

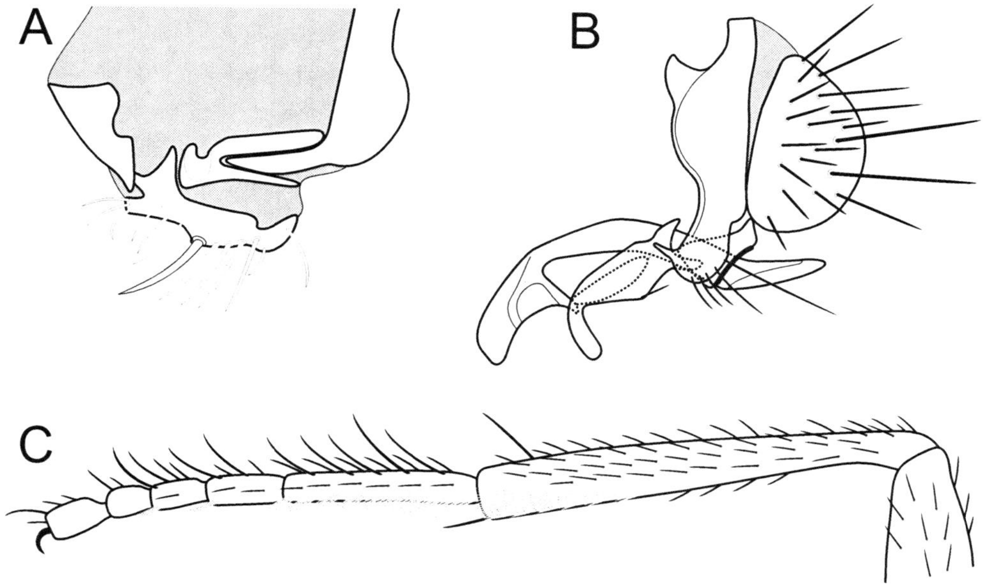

Fig. 18. *Drosophila albifacies* male, Waikamoi specimen. (A) Lateral view of labellum. (B) Terminalia. (C) Right front leg, anterior view.

2. (1) Face white. Hawai'i, Maui... *D. albifacies* Hardy
— Face brown..3.

3. (2) Katepisternum brown; front basitarsus lacking ciliation except at the apex. Maui.. *D. curticilia* Hardy
— Katepisternum yellow; front basitarsus with 5–6 cilia along the segment. O'ahu .... *D. dolomata* Hardy

Drosophila albifacies Hardy **new island record**
Fig. 18

A single specimen from Maui that appears to be *D. albifacies* is in the UHIM (Waikamoi, S1, 10.xi.1968, JPM). This species was previously known only from Hawai'i. The Maui specimen has a prominent thumb-like projection on the anterodorsal margin of the epandrial lobe that is not shown in the original description (Fig. 18B). However, another white-faced specimen from Hawai'i (Hualālai, Moanuiahea, 3400 ft., M86B, 23.xii.1969, KYK) does have it. While it is possible that there may be a Hawai'i endemic species and one found on both Hawai'i and Maui, the area of the epandrium anterior of the antecosta (acrotergite) is clearly variable in other species, sometimes differing between the two sides in an individual. Therefore it seems more prudent to keep all specimens under *D. albifacies* until more specimens are available for comparison.

scolostoma subgroup

Species in this subgroup are rarely collected and poorly known, and the restricted distribution suggests that undescribed species may exist. It may also represent a morphologically divergent offshoot of another lineage, but there are no clear relationships with other subgroups. The only rearing record, for *D. deltaneuron*, is from *Pritchardia* fruit, a highly unusual host (Magnacca et al. 2008).

1. Dm–cu crossvein forked anteriorly; wings infuscated at the apex, crossveins, and cell c. O'ahu ....*D. deltaneuron* Bryan

— Dm–cu crossvein normal; wings subhyaline. Maui.................................2.

2. (1) Front basitarsus with dorsal cilia extending the entire length ....*D. scolostoma* Hardy

— Front basitarsus with cilia only on the apical half..........*D. mediana* Hardy

semifuscata subgroup

With transfer of *D. dolomata* and its relatives to the *nanella* subgroup, the mouthparts of the remaining members of the *bridwelli* and *semifuscata* subgroups are virtually identical. The only character separating the two subgroups, infuscation of the anterior margin of the wing, is often vague and inconsistent. Therefore, the *bridwelli* and *semifuscata* subgroups are combined under the latter name. In addition, *D. sadleria* is added to the subgroup and *D. acrostichalis* is removed from it as an unplaced species (see Discussion under the respective species). Earlier classifications relied heavily on wing infuscation, which led to unrelated species such as *D. humeralis* (as *D. aethostoma*) being placed in this subgroup (Hardy and Kaneshiro 1968). The degree of infuscation, particularly along the anterior margin of the wing, is highly variable in some species, such as *D. xuthoptera*.

This subgroup is separable into two complexes, which may or may not be monophyletic: dark-bodied species (*D. acanthostoma*, *D. anoplostoma*, *D. mandibulata*, *D. olaae*, *D. peloristoma*, and *D. semifuscata*) and pale-bodied species (*D. apicipuncta*, *D. bridwelli*, *D. diminuens*, *D. magnimacula*, *D. sadleria*, *D. wawae*, *D. xuthoptera*, and *D. z-notata*). The species of the former complex with rearing records all breed in sap fluxes of various trees, while the only records for the latter (*D. apicipuncta* and *D. sadleria*) are for mining in live *Sadleria* fern rachis (Magnacca et al. 2008).

Pinned specimens of this subgroup often have the labellar lobes appressed, concealing the long curved spines in the folds of dried membrane. At least three, *D. anoplostoma*, *D. peloristoma*, and *D. semifuscata*, appear to have lost their spinose setae. Nearly all species are rarely collected, and most are known from very few specimens or collections.

1. Mesonotum brown to black, pleura usually similar to mesonotum; if pleura light brown, then the labellum lacks spine-like setae ....2.
— Mesonotum yellow to rufous (with darker stripes in *D. sadleria*), pleura usually with both light and dark areas; labellum always with dorsal spine-like setae ....7.

2. (1) Front tarsus with cilia only on the basitarsus (Fig. 19C); wing entirely hyaline. Maui ....*D. mandibulata* n. sp.
— Front tarsus with cilia on at least the first two segments; wing infuscated on the dm–cu crossvein and/or anterior margin, though sometimes faint......3.

3. (2) Wing with a distinct mark only on the dm–cu crossvein; front tarsus with long cilia on segments 1–4; palpi of normal form, relatively narrow, broadest before the apex, rounded apically. Hawai'i, Maui, Moloka'i .... *D. olaae* Grimshaw
— Entire anterior margin of wing infuscated, or if faint, then the dm–cu crossvein is not much darker; front tarsus with long cilia only on the first three segments; palpi triangular, broadest at the apex, apex truncate, transverse or slightly emarginate ....4.

4. (3) Labellar sclerite more or less quadrate, lacking elongate spines; palpi yellow; femora not more than faintly tinged with brown; thorax moderately dark brown, humeri paler..5.
— Labellar sclerite expanded, semicircular, spines variable; palpi brown; femora distinctly dark brown; thorax nearly black, humeri not paler ....6.

5. (4) Pleura almost entirely brown. O'ahu ....*D. peloristoma* n. sp.
— Pleura largely light brown, with darker markings especially at the upper margin of the mesopleuron. Maui............................*D. semifuscata* Hardy

6. (4) Labellum with long, curved spines dorsally. Hawai'i, Maui? ....*D. acanthostoma* Hardy and Kaneshiro
— Labellum lacking long curved spines. Maui, Moloka'i, Lāna'i, Hawai'i?.. ....*D. anoplostoma* Hardy and Kaneshiro

7. (1) Wing with a large, conspicuous mark on the r–m crossvein, about the same size as that on the dm–cu crossvein. O'ahu ....*D. magnimacula* Hardy
— If a mark on the r–m crossvein is present, it is not much larger than the vein and not particularly conspicuous ....8.

8. (7) Front tarsus with elongate cilia on at least the first three segments...........9.
— Cilia present only on the first two segments of the front tarsus (sometimes a single long cilium on the third segment) ....11.

9. (8) Wing hyaline except for the dm–cu crossvein. O‘ahu .... *D. bridwelli* Hardy
— Wing with more extensive markings, at the apex and/or anterior margin (sometimes diffuse).. 10.

10. (9) Labellar spines short (Fig. 21A); anterior wing margin infuscated. Hawai‘i .... *D. wawae* n. sp.
— Labellar spines long (as in Fig. 19A); wing with a faint basal infuscation from cell c to the r–m crossvein but not continuing across the anterior margin. Hawai‘i .... *D. diminuens* Hardy

11. (8) Wing margin infuscated between apices of veins M and CuA_1; second segment of front tarsus with only 2 long cilia. O‘ahu ....*D. z-notata* Bryan
— Wing margin hyaline between apices of veins M and CuA_1; second segment of front tarsus with at least 3 long cilia .... 12.

12. (11) Front basitarsus with elongate cilia over only the apical 1/4–1/3. Hawai‘i. ....*D. apicipuncta* Hardy
— Front basitarsus with elongate cilia extending over at least the apical half.. .... 13.

13. (12) Anterior wing margin not infuscated; anterior reclinate seta about midway between proclinate and posterior reclinate; costal fringe extending over halfway between the apices of R_{2+3} and R_{4+5}. O‘ahu.......*D. sadleria* Bryan
— Anterior wing margin usually completely infuscated, sometimes faint or absent; anterior reclinate seta only slightly posterior of the proclinate, their bases almost overlapping; costal fringe extending only 1/3–2/5 the distance between the apices of R_{2+3} and R_{4+5}. Maui, Moloka‘i .... *D. xuthoptera* Hardy

Drosophila acanthostoma Hardy and Kaneshiro

Drosophila acanthostoma Hardy and Kaneshiro, 1968:244–247. Holotype ♂, Hawai‘i, forest above Pa‘auilo, 3200 ft., collected on trunk of *Myrica faya*, 24.viii.1965, KYK. BPBM 8924 [examined].

Diagnosis. Dark-bodied species with the anterior half of the wing infuscated; united with *D. anoplostoma* and distinguished from other dark-bodied members of the subgroup by the prominent, semicircular labellar sclerite and the very broad, apically truncate or emarginate palpi. Separated from *D. anoplostoma* by the presence of elongate, curved spines at the dorsal end of the labellum, similar to those found in most of the pale-bodied species.

Description. ♂. **Head.** Front dark brown to black, pollinose; fronto-orbital plates and ocellar triangle similar, sometimes appearing darker. Anterior reclinate seta about even with the proclinate, about 2/3–3/4 as long; posterior reclinate nearly 2.5 times as long as the anterior. Ocellar setae about as long as the posterior reclinate. Vertical setae normal in position, slightly longer than the posterior reclinate. Eyes with short, inconspicuous pile, about as long as one facet. Face brown, with a low median carina. Antenna brown; arista with 5–7 dorsal and 2–3 ventral rays in addition to the apical fork. Two strong setae present at the vibrissal angle, hardly differentiated from the posterior oral setulae, which are also relatively long. Gena brown; about as wide as 3 eye facets. Palp brown, very broad, flattened, setulose, almost triangular, broadest at the apex which is truncate or slightly emarginate; with a moderately strong apical seta on the outer corner and a subapical seta along the outer margin just basal of it, both of these stronger but not much longer than the setulae. Labellar sclerite semicircular, fringed with hair-like setae and dorsally with a cluster of 3 elongate, closely-placed, inclinate spines. **Thorax.** Entirely dark brown to black and pollinose, except humerus sometimes tinged with yellow. Two humeral setae, the ventral nearly as long as the dorsal. Acrostichal setulae in about 6–10 irregular rows. Two pairs of dorsocentral setae, the anterior about 2/3–3/4 as long as the posterior. Halteres yellow, stems brown. **Legs.** Coxae and femora brown, tibiae and tarsi yellow. Front tibia lacking ciliation. Front basitarsus with three rows of elongate cilia along the entire length, with 4 in the posterodorsal and 6–7 each in the dorsal and anterodorsal rows; second segment continuing these with 1, 2, and 2 cilia respectively; third segment with a pair of elongate cilia at the apex; remaining segments without long cilia. **Wings.** Entirely infuscated anterior of R_{4+5}, as well as over the dm–cu crossvein, basally along M to the crossvein, and apical segment of CuA_1; markings diffuse, sometimes faint. Costal fringe extending about 1/2–3/5 the distance between the apex of R_{2+3} and R_{4+5}. **Abdomen.** Entirely brown. Epandrium almost parallel-sided, narrowing slightly at the dorsal margin. Aedeagal apodeme about twice as long as high; ventral margin slightly sinuate, not strongly concave. Aedeagus strongly bent, basal angle acute, apex elongate; preapical protuberance small, low.

♀. Identical to the male with the following exceptions. **Head.** Apex of palp rounded, not truncate, with a single strong, subapical seta. Labellum unmodified. **Legs.** Tibiae tinged with brown. Front legs without cilia. **Abdomen.** Ovipositor broadly triangular, apex acute.

Material Examined. Hawai‘i: 1♀, allotype, same data as holotype, G6.1, 25.viii.1965, KYK (BPBM). 1♂ paratype, forest above Pa‘auilo, 3000 ft., C143, 26.vii.1965, HLC. 2♂ paratypes, forest above Pa‘auilo, 3200 ft., collected on *Myrica* [=*Morella*] *faya*, 6.viii.1965, DEH. 1♂, same date and locality, DEH. 2♂ 3♀, Pu‘u Wa‘awa‘a, 4000 ft., ex *Osmanthus* [=*Nestegis*] sap flux, 1.viii.1971, SLM. 1♂, same date and locality, 3800 ft., sweeping under *Pisonia* grove, SLM. 2♂ 1♀, Pāpā, ex ‘ōhi‘a [=*Metrosideros polymorpha*] sap flux, 3.x.1973, SLM. 1♀, Greenwell Ranch, Pauahi, site B, T90, 24.x.1975, HTS. 3♂ 5♀, Pu‘u Wa‘awa‘a, ex ‘ōhi‘a sap flux, 8.viii.1978, SLM. Maui: 1♀ paratype[?], Auwahi, 3000 ft., 17.vi.1965, DEH. 1♂ paratype[?], Auwahi, G7.5, 27.viii.1965, KYK (all except allotype at UHIM).

Distribution & Ecology. Hawai‘i, Maui?; known from scattered mesic forest sites. Reared from sap flux of *Metrosideros polymorpha* (‘ōhi‘a, Myrtaceae) and *Nestegis sandwicensis* (olopua, Oleaceae; Magnacca and O'Grady 2008).

Discussion. The male mouthparts, front leg, and genitalia, and ovipositor were illustrated by Hardy and Kaneshiro (1968, Fig. 31). The description states that the Maui specimens were not to be designated as paratypes, but those at UHIM bear paratype labels. The distribution of this species and its sister species is uncertain; see Discussion under *D. anoplostoma*.

Drosophila anoplostoma Hardy and Kaneshiro

Drosophila anoplostoma Hardy and Kaneshiro, 1968:250. Holotype ♂, Maui, Auwahi, 3700 ft., J10G, 25.vii.1966, HLC. BPBM 8926 [examined].

Diagnosis. See Description.

Description. Identical to *D. acanthostoma* with the following exceptions. ♂. **Head.** Labellum lacking elongate spine-like setae. **Legs.** All tibiae distinctly brown, though not as dark as femora. Front basitarsus with only 3–4 cilia in the dorsal (middle) row and 2 in the posterodorsal row.
Material Examined. Maui: 1♀, allotype, Auwahi, 3700 ft., 18.vi.1965, KYK (BPBM). 1♂ paratype, same data as holotype. 1♂, Auwahi, 5–7.viii.1964, HLC. 9♂, Kaupō, 4000 ft., ex *Acacia koa* sap flux, 2.iv.1971, SLM. 1♂, Olowalu, 1500 ft., ex *Osmanthus* [=*Nestegis*] sap flux, 19.iv.1971, SLM. 6♂ 3♀, Kaupō Gap, ex *Myoporum* sap flux, 20–21.iv.1971, SLM. 1♀, Manawainui Gulch, R11, 29.iv.1972, KYK. Moloka‘i: 1♂ 1♀, Kawela Gulch, ex *Osmanthus* flux, 11.ii.1972, SLM. Lāna‘i: 4♂ 7♀, ‘Āwehi, 2800 ft., Q1, reared ex *Osmanthus* flux, 4.vi.1971, SLM. Hawai‘i: 1♂, Pa‘auilo, C143.5, 26.viii.1965, HLC (all except allotype at UHIM).

Distribution & Ecology. Maui, Moloka‘i, Lāna‘i, Hawai‘i; mesic forest, generally on south-facing slopes. Reared from sap flux of *Acacia koa* (koa, Fabaceae), *Metrosideros polymorpha* (‘ōhi‘a, Myrtaceae), *Myoporum sandwicense* (naio, Myoporaceae) and *Nestegis sandwicensis* (olopua, Oleaceae; Magnacca and O'Grady 2008).

Discussion. The male mouthparts, front leg, and genitalia were illustrated by Hardy and Kaneshiro (1968, Fig. 33). *Drosophila acanthostoma* and *D. anoplostoma* are very similar species, apparently separable only by the three characters listed above. In general, specimens from the island of Hawai‘i have labellar spines and more basitarsal cilia, while those from Maui lack spines and have fewer cilia. However, the presence of both spined and spineless morphotypes at Pa'auilo (Hawai'i) and Auwahi (Maui) suggests the possibility that these are variants and not distinct species. Some may have had the spines broken off, but one *D. anoplostoma* paratype has been dissected and clearly is genuinely lacking spines. The mouthpart and tarsal cilia characters appear to coincide in all specimens where both are clearly visible, but the cilia are often difficult to count and clearly exhibit some variation within each mouthpart morphotype. Further work, including DNA analysis and more thorough morphological study, is necessary to determine if these populations consist of two multi-island species with different mouthpart morphologies; two species with similar variation on Maui and Hawai'i; four species, one for each morphotype on each island; or a single species. Unfortunately no recent specimens are available to conduct such a study.

Drosophila apicipuncta Hardy

Drosophila apicipuncta Hardy, 1965:150–152. Holotype ♂, Hawai'i, Upper Hamakua Ditch Trail, on miscellaneous ferns, 10.iv.1929, OHS. BPBM 6294 [examined].

Diagnosis. A member of the pale-bodied species complex, with spinose mouthparts and small wing marks; similar to and sympatric with *D. diminuens*, distinguished by the reduced ciliation of the front tarsus.

Description. ♂. **Head.** Front brown above, yellow near frontal suture; pollinose. Anterior reclinate seta slightly posterior of the proclinate, 2/3–3/4 as long; posterior reclinate 2.5–3 times as long as the anterior. Ocellar setae about as long as the posterior reclinate. Vertical setae normal in position, about as long as the posterior reclinate. Eyes with short, inconspicuous pile, about as long as one facet. Face yellow-brown, darker ventrally, with a rounded median carina. Antenna brown, yellow at the base of segment 3; arista with 6–7 dorsal and 2–3 ventral rays in addition to the apical fork. Several elongate setae present at the vibrissal angle, these not differentiated from the other oral setulae, which are also relatively long. Gena yellow, narrowly brown along the oral margin; about as wide as 4 eye facets. Palp yellow, flattened, setulose, broadest before the apex which is rounded; with a single strong apical seta. Labellar sclerite with 2 closely-placed strong, brown, curved, inclinate, round, spine-like setae at the dorsal end (often appearing as one), followed by 2–3 shorter and weaker spines; ventral half with about 8–10 hair-like setae. **Thorax.** Mesonotum yellow, tinged faintly with brown; pleura yellow with two brown stripes, one each at the dorsal and ventral margins of the anepisternum and anepimeron. Two humeral setae, the ventral nearly as long as the dorsal. Acrostichal setulae in about 6–10 irregular rows. Two pairs of dorsocentral setae, the anterior about 2/3–3/4 as long as the posterior. Halteres yellow, stems brown. **Legs.** Yellow. Front tibia lacking ciliation. Front basitarsus with about 7 elongate

cilia over the apical third of the dorsal and anterodorsal surfaces, not in distinct rows; second segment with 5–7 similar cilia; remaining segments without long cilia. **Wings.** Largely hyaline; with infuscations over the dm–cu crossvein and at the apices of R_{2+3}, R_{4+5}, and M. Costal fringe extending about 1/3 the distance between the apex of R_{2+3} and R_{4+5}. **Abdomen.** Each segment dark brown posteriorly, with a distinct yellow band (sometimes medially interrupted) across the anterior half of each segment.
♀. Identical to the male with the following exceptions. **Head.** Labellum unmodified. **Legs.** Front legs without cilia. **Abdomen.** Ovipositor broadly triangular, apex blunt.

MATERIAL EXAMINED. HAWAI'I: 1♀, allotype, Kīlauea, 24.vi.1934, reared from rachis of *Sadleria* fern, OHS (BPBM). 3♂, Kīlauea Field Station, sweeping *Sadleria*, 22.vi.2006, KNM (UCPO).

DISTRIBUTION & ECOLOGY. Hawai'i. Reared from mines in living *Sadleria* ('ama'u, Blechnaceae) fern rachises (Bryan 1938; Hardy 1965).

DISCUSSION. The male front leg, wing, and genitalia, and ovipositor were illustrated by Hardy (1965, Fig. 32). Individuals of this species were originally determined by Bryan (1938) as *D. sadleria*, which is of similar appearance and shares the same breeding habitat on O'ahu. The males of *D. apicipuncta* and *D. diminuens* can be distinguished by the ciliation of the front legs, but it is uncertain whether the females can be separated. From the specimens on hand it appears that *D. apicipuncta* possesses fewer lateral ovisensillae on the ovipositor than *D. diminuens* (4–5 vs. 9), but only two specimens of each were available for comparison and the range of variation in this character is not known.

Drosophila bridwelli Hardy

Drosophila bridwelli Hardy, 1965:189–191. Holotype ♂, O'ahu, Hau'ula, 2.viii.1914 [no collector]. BPBM 6314 [examined].

DIAGNOSIS. Similar to *D. apicipuncta* and related species; separated by lacking wing marks except over the dm–cu crossvein and by having cilia on segments 1–4 of the front tarsus.

DESCRIPTION. ♂. **Head.** Front brownish yellow, dull; fronto-orbital plates and ocellar triangle darker, brown. Anterior reclinate seta slightly posterior of the proclinate, about equal in length. Vertical setae normal in position. Eyes with short, inconspicuous pile, about as long as one facet. Face yellow-brown, darker ventrally, with a rounded median carina which is paler. Antenna yellow, tinged with brown; arista with 7 dorsal and 3 ventral rays in addition to the apical fork. No strong oral vibrissae present. Gena yellow, narrowly brown along the oral margin; about as wide as 4 eye facets. Palp yellow, flattened, setulose, broadest before the

apex which is rounded; with a single strong apical seta. Labellar sclerite with 2–3 closely-placed strong, brown, curved, inclinate, round, spine-like setae at the dorsal end (often appearing as one), followed by 2 shorter and weaker spines; ventral half with numerous hair-like setae. **Thorax.** Mesonotum yellow to rufous, tinged faintly with brown; pleura yellow with two brown stripes, at the dorsal and ventral margins of the anepisternum and anepimeron. Two humeral setae, the ventral nearly as long as the dorsal. Acrostichal setulae in about 8 irregular rows. Two pairs of dorsocentral setae. **Legs.** Yellow. Front tibia lacking ciliation. Front basitarsus with about 6 elongate dorsal cilia over the apical half, continuing on segments 2–4. **Wings.** Largely hyaline, infuscated only over the dm–cu crossvein. Costal fringe extending about half the distance between the apex of R_{2+3} and R_{4+5}. **Abdomen.** Entirely dark brown. Epandrium parallel-sided, not narrowed dorsally.
♀. Identical to the male with the following exceptions. **Head.** Three to four strong oral vibrissae present. Labellum unmodified. **Legs.** Front legs without cilia. **Abdomen.** Ovipositor broadly triangular, apex blunt.

Material Examined. O'ahu: 1♂, Kahana, 7.ix.1924, OHS (UHIM).

Distribution & Ecology. O'ahu. Breeding habits unknown.

Discussion. The male antenna, mouthparts, front leg, and genitalia, and ovipositor were illustrated by Hardy (1965, Fig. 51).

Drosophila diminuens Hardy

Drosophila diminuens Hardy, 1965:244–245. Holotype ♂, Hawai'i, Kīlauea, 22.iv.1920, OHS. BPBM 6341 [examined].

Diagnosis. Similar to *D. apicipuncta*: labellum with elongate curved spines, body yellow-brown, wing marks at the apices of the veins and crossveins. Distinguished by having cilia on the first three segments of the front tarsus rather than only the first two.

Description. ♂. **Head.** Front brown above, yellow near frontal suture; pollinose. Anterior reclinate seta slightly posterior of the proclinate, 2/3–3/4 as long; posterior reclinate 2.5–3 times as long as the anterior. Ocellar setae about as long as the posterior reclinate. Vertical setae normal in position, about as long as the posterior reclinate. Eyes with short, inconspicuous pile, about as long as one facet. Face yellow-brown, darker ventrally, with a rounded median carina. Antenna brown, yellow at the base of segment 3; arista with 6–7 dorsal and 2–3 ventral rays in addition to the apical fork. Several elongate setae present at the vibrissal angle, these not differentiated from the other oral setulae, which are also relatively long. Gena yellow, narrowly brown along the oral margin; about as wide as 4 eye facets. Palp yellow, flattened, setulose, broadest before the apex which is rounded; with a single strong apical seta. Labellar sclerite with 2 closely-placed strong, brown,

curved, inclinate, round, spine-like setae at the dorsal end (often appearing as one), followed by 2–3 shorter and weaker spines; ventral half with about 8–10 hair-like setae. **Thorax.** Mesonotum yellow, tinged faintly with brown; pleura yellow with two brown stripes, one each at the dorsal and ventral margins of the anepisternum and anepimeron. Two humeral setae, the ventral nearly as long as the dorsal. Acrostichal setulae in about 6–10 irregular rows. Two pairs of dorsocentral setae, the anterior about 2/3–3/4 as long as the posterior. Halteres yellow, stems brown. **Legs.** Yellow. Front tibia lacking ciliation. Front basitarsus with about 12 elongate cilia over the apical half of the dorsal and anterodorsal surfaces, not in distinct rows; second segment with 5–7 similar cilia; third segment with 2–3 elongate dorsal cilia; remaining segments without long cilia. **Wings.** Largely hyaline; with infuscations over the dm–cu crossvein, at the base of R_{4+5} to the r–m crossvein (often faint, inconspicuous, or absent), and at the apices of R_{2+3}, R_{4+5}, and M. Costal fringe extending about 1/3 the distance between the apex of R_{2+3} and R_{4+5}. **Abdomen.** Each segment dark brown posteriorly, with a distinct yellow band (sometimes medially interrupted) across the anterior half of each segment in fresh specimens (often discoloring so that the entire abdomen appears dark).
♀. Identical to the male with the following exceptions. **Head.** Three to four strong oral vibrissae present. Labellum unmodified. **Legs.** Front legs without cilia. **Abdomen.** Ovipositor broadly triangular, apex blunt.

MATERIAL EXAMINED. HAWAIʻI: 1♂ paratype, Kīlauea, ex light trap, viii.1958, JWB. 1♂, Puʻu Laʻalaʻau, Kohala Mts., 3500 ft., 29.viii.1966, KYK. 1♂, Saddle Rd., Kīpuka #2, 5400 ft., K34, 10.viii.1967, WBH. 1♂, Keaʻau, ʻŌlaʻa [note: these are separate, non-overlapping localities], 2300 ft., K55, 29.ix.1967, HLC. 2♂, Kīlauea, 10.iv.1969, G. K. Kobayashi (all above at UHIM). 1♂, Stainback Highway, R Road, 3200 ft., 1.x.2006, KNM (UCPO). 1♂, Saddle Road, Kīpuka #9, 5100 ft., 3.x.2006, KNM (UCPO).

DISTRIBUTION & ECOLOGY. Hawaiʻi. Breeding habits unknown; probably mining in *Sadleria* rachises like *D. apicipuncta*.

DISCUSSION. The male mouthparts, front leg, and wing were illustrated by Hardy (1965, Fig. 79). See Discussion under *D. apicipuncta* regarding separation of females.

Drosophila magnimacula Hardy

Drosophila magnimacula Hardy, 1965:350–352. Holotype ♂, Oʻahu, Pūpūkea Trail, vii.1958, in banana thicket, DEH. BPBM 6393 [examined; see Discussion].

DIAGNOSIS. Belonging to the *D. apicipuncta* complex of pale-bodied species with spinose mouthparts; easily distinguished by the prominent wing marks, including the unusually large one over the r–m crossvein, as well as the front tarsus with cilia on the first 4 segments and long costal fringe.

DESCRIPTION. ♂. **Head.** Front yellow, tinged with brown near vertex lightly pollinose; ocellar triangle and fronto-orbital plates brownish. Anterior reclinate seta slightly posterior of the proclinate, 2/3–3/4 as long; posterior reclinate about twice as long as the anterior. Ocellar setae about as long as the posterior reclinate. Vertical setae normal in position, slightly longer than the posterior reclinate. Eyes with short, inconspicuous pile, about as long as one facet. Face brown, with a rounded median carina. Antenna pale yellowish brown, yellow at the base of segment 3; arista with 6–8 dorsal and 3–4 ventral rays in addition to the apical fork. No oral vibrissae present. Gena yellow, narrowly brown along the oral margin; about as wide as 3 eye facets. Palp yellow, flattened, setulose, broadest before the apex which is rounded; with a single strong apical seta. Labellum with 2 closely-placed strong, brown, curved, inclinate, round, spine-like setae at the dorsal end, appressed and appearing as one, followed by 2–3 slightly shorter, more widely spaced spines; ventral half with numerous hair-like setae. **Thorax.** Yellow, mesonotum faintly tinged with brown; pleura faintly marked with brown along dorsal margin of anepisternum and on anepimeron. Two humeral setae, the ventral nearly as long as the dorsal. Acrostichal setulae in about 6–10 irregular rows. Two pairs of dorsocentral setae, the anterior about 3/4 as long as the posterior. Halteres yellow, stems brown. **Legs.** Yellow. Front tibia lacking ciliation. Front basitarsus with 4 elongate cilia each in a dorsal and anterodorsal row over the apical 2/3; second segment with a dorsal and an anterodorsal cilia near the middle; second, third, and fourth segments each with 3 elongate dorsal cilia in a transverse row at the apex. **Wings.** Hyaline with large, conspicuous infuscations at the apices of R_{2+3}, R_{4+5}, and M, and over the dm–cu and r–m crossveins; the last unusually large, about the same size as the mark on the dm–cu crossvein. Costal fringe long, extending about 2/3 the distance between the apex of R_{2+3} and R_{4+5}. **Abdomen.** Entirely brown. Aedeagal apex elongate, acute; preapical protuberance small, low.
♀. Identical to the male with the following exceptions. **Head.** Labellum unmodified. **Legs.** Front legs without cilia. **Abdomen.** Ovipositor broadly triangular, apex blunt.

MATERIAL EXAMINED. O'AHU: 1♀, allotype, Lulumahu Valley, banana stem, 14.vi.1938, FXW (BPBM). 1♀, paratype, Mt. Olympus, 18.vi.1918 (BMNH). 1♀, Pūpūkea, 4.ii.1964, DG (BMNH). 1♂, Pūpūkea, 14.vi.1964, M. R. Wheeler (UHIM). 1♂, same locality and collector, 11.vii.1964 (UHIM).

DISTRIBUTION & ECOLOGY. O'ahu. Breeding habits unknown.

DISCUSSION. The male mouthparts, front tarsus, wing, and genitalia were illustrated by Hardy (1965, Fig. 131). The holotype of this species has been the subject of some confusion. A specimen labelled with the correct data for the holotype is present in the BPBM type collection, but it bears a paratype label; no other specimens with this data were noted by Hardy (1965). The original description also states that there were seven paratypes, including one that was deposited at BPBM, but does not say which specimens went to which institutions. We have only been able to locate six (none at BPBM), which account for all the localities and dates given by Hardy (1965). In the catalog of Bishop Museum entomological types, Evenhuis (1982) lists *D. magnimacula* as "HOLOTYPE destroyed, pin and labels missing;" however, this may have been based on the assumption that the Pūpūkea, 1958 specimen was intended as a paratype. Given that there was only supposed to be one specimen from Pūpūkea, 1958 and it is still present, it is undoubtedly the intended holotype, but inadvertently received a paratype label and was included in the count of paratypes. The specimen has been now been given a holotype label.

Drosophila mandibulata **n. sp.**
Fig. 19

DIAGNOSIS. A dark-bodied species with prominent labellar spines. Easily distinguished from all other members of the subgroup by the completely hyaline wings and by having long cilia only on the first segment of the front tarsus.

DESCRIPTION. ♂. **Head.** Front brown, darker and more pollinose on fronto-orbital plates and ocellar triangle. Anterior reclinate seta slightly posterior of the proclinate, about half as long; posterior reclinate about 2.5 times as long as the anterior. Ocellar setae about as long as the posterior reclinate. Vertical setae normal in position, slightly longer than the posterior reclinate. Eyes with short, inconspicuous pile, about as long as one facet. Face brown, paler and with a rounded carina dorsomedially. Antenna dark brown, paler on the apicodorsal surface of the second segment; arista with 6–8 dorsal and 3–4 ventral rays in addition to the apical fork. Several elongate oral setae present at the vibrissal angle (slightly longer than the posterior oral setulae, which are also relatively long), but no distinctly stronger oral vibrissae. Gena yellow, narrowly brown along the oral margin; about as wide as 5–6 eye facets. Palp brown, paler towards the base, flattened, broadest before the apex, setulose; with an elongate subapical seta. Labellum with 2 closely-placed strong, brown, curved, inclinate, round, spine-like setae at the dorsal end, followed by 3 shorter and weaker spines; ventral half with about 8–10 hair-like setae (Fig. 19A). **Thorax.** Entirely brown, pollinose. Two humeral setae, the ventral about as long as the dorsal. Acrostichal setulae in about 8 irregular rows. Two pairs of dorsocentral setae, the anterior about 2/3 as long as the posterior. Halteres pale brown, stems darker. **Legs.** Brown. Front tibia lacking ciliation. Front basitarsus with 4–6 elongate cilia each in irregular dorsal and anterodorsal rows; remaining segments lacking long cilia (Fig. 19C). **Wings.** Entirely hyaline, faintly yellowish. Costal fringe extending about 1/3 the distance between the apex of R_{2+3} and R_{4+5}. **Abdomen.** Entirely dark brown. Epandrium slightly narrowed dorsally; anteroventral spine curled under, thus appearing narrow in lateral view. Aedeagal apodeme about twice as long as high, strongly concave

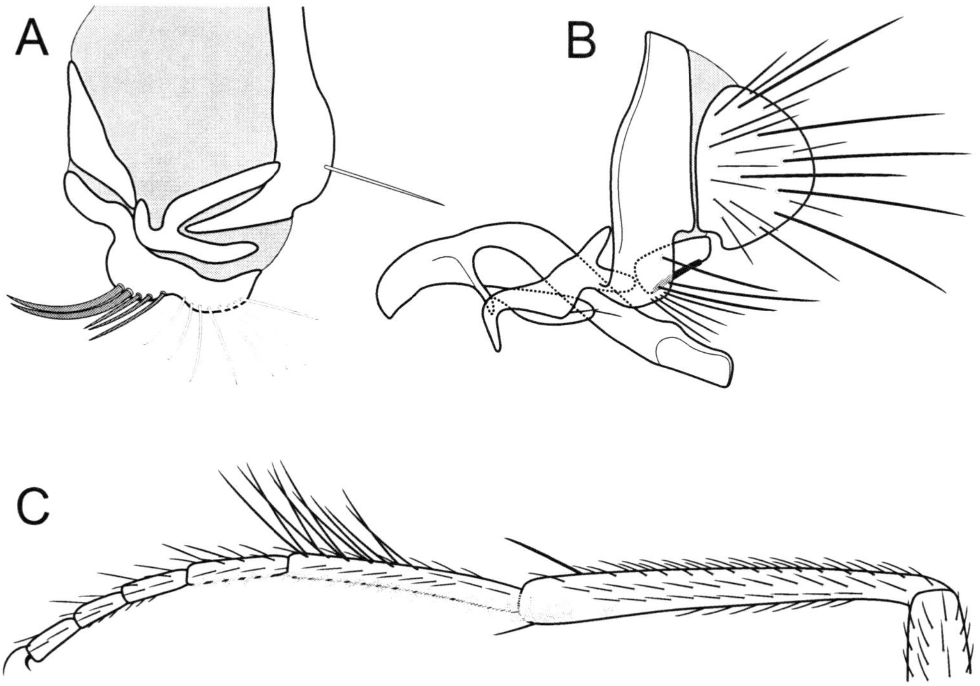

Fig. 19. *Drosophila mandibulata* male. (A) Lateral view of labellum. (B) Terminalia. (C) Right front leg, anterior view.

ventrally. Aedeagus narrow at the base, broader beyond, strongly bent, basal angle acute, apex quadrate and blunt at the tip; preapical protuberance small, low (Fig. 19B).

♀. Unknown.

MEASUREMENTS. N = 1♂. TL = 1.82 mm; WL = 4.22 mm; TL/WL = 0.4; HW = 1.32 mm; HW/FS = 2.1; HW/TL = 0.7; CI = 4.8; 4V = 1.2; 5X = 1.2; 4C = 0.5; M = 0.3.

TYPES. MAUI: Holotype ♂ (BPBM 16737), Waikamoi, 4300 ft., 9.vii.1964, WBH.

PARATYPES. MAUI: 1♂, Waikamoi, 4000 ft., 24.vii.1965, DEH (UHIM). MOLOKA'I: 1♂, Pu'u Haha, 16.vii.1963, LHT (UHIM).

DISTRIBUTION & ECOLOGY. Maui and Moloka'i. Breeding hosts unknown, but resembles species that breed in sap flux.

ETYMOLOGY. From the Latin *mandibula*, jaw, referring to the resemblance of the labellar spines to insect mandibles.

Drosophila olaae Grimshaw

Drosophila olaae Grimshaw, 1901:66. Holotype ♀, Hawai'i, 'Ōla'a, xii.1896, RCLP. BMNH [examined].

Diagnosis. Similar to *D. mandibulata*, a dark-bodied species with the anterior margin of the wing completely hyaline; but with a distinct mark over the dm-cu crossvein, and front tarsus with cilia on segments 1–4.

Description. ♂. **Head.** Front reddish brown, slightly paler anteriorly, pollinose; fronto-orbital plates and ocellar triangle dark brown and more densely pollinose. Anterior reclinate seta slightly above the proclinate, about equal in length but much weaker; posterior reclinate about twice as long as the anterior. Ocellar setae about as long as the posterior reclinate. Vertical setae normal in position, slightly longer than the posterior reclinate. Eyes with short, inconspicuous pile, about as long as one facet. Face brown, with a low median carina. Antenna largely brown, yellow on apex and inner surface of second segment and extreme base of third; arista with 7–8 dorsal and 3–4 ventral rays in addition to the apical fork. Vibrissal angle with only very small setulae, other oral setae becoming stronger posteriorly. Gena yellowish brown; about as wide as 4 eye facets. Palp brown, paler at base, flattened, setulose, broadest before the apex which is rounded; with a strong, elongate apical seta. Labellar sclerite narrow, with 2 closely-placed strong, brown, nearly straight, slightly inclinate, round, spine-like setae at the dorsal end, appressed and appearing as one, closely followed by 2 slightly shorter spines; ventral half with several hair-like setae. **Thorax.** Entirely dark brown to black, tinged with rufous, pollinose. Two humeral setae, the ventral nearly as long as the dorsal. Acrostichal setulae in 8–10 irregular rows. Two pairs of dorsocentral setae, the anterior about 3/4 as long as the posterior. Halteres yellow, stems brown. **Legs.** Coxae and femora brown, tibiae and tarsi brownish yellow; apex of each tibia with a brown band. Front tibia lacking ciliation, but entire posterior surface covered with short erect setulae. Front basitarsus with 8–9 strong, elongate cilia in an anterodorsal row and 5–6 thinner cilia in a dorsal row over the apical 3/4; second segment with a dorsal and an anterodorsal cilia near the middle; second, third, and fourth segments each with 3 elongate dorsal cilia in a transverse row at the apex; posterior surface of all segments with numerous short, erect setulae, sometimes with a few longer emergent cilia. **Wings.** Mostly hyaline, conspicuously infuscated over the dm–cu crossvein; sometimes bearing very faint infuscations at the apices of R_{2+3}, R_{4+5}, and M. Costal fringe extending about 1/2–3/5 the distance between the apex of R_{2+3} and R_{4+5}. **Abdomen.** Almost entirely dark brown, posterior margins of tergites narrowly yellow, sometimes with indistinct anterolateral spots. Epandrium almost parallel-sided, not narrowing dorsally.

♀. Identical to the male with the following exceptions. **Head.** Two or three long setae present at the vibrissal angle, not strongly differentiated from the other oral setae. Labellum unmodified. **Legs.** Tibiae appear to be all brown (may be discolored). Front legs without cilia. **Abdomen.** Ovipositor broadly triangular, apex blunt.

MATERIAL EXAMINED. HAWAI'I: 1♂, Upper 'Ōla'a Forest, viii.1952, labelled "homotype," DEH. 1♂, 'Ōla'a Forest Reserve, 3775 ft., G9.6, 3.ix.1965, KYK. MAUI: 1♂ 1♀, Haleakalā, 1500 ft. [no date or collector]. 1♂, Haleakalā, 2000 ft. [no date or collector]. MOLOKA'I: 1♀, Pu'u Kolekole, 3600 ft., vii.1953, DEH. 1♂, 'Ape'e, 1700 ft., L97, 23–25.vii.1968, HLC (all UHIM).

DISTRIBUTION & ECOLOGY. Hawai'i, Maui, and Moloka'i. Breeding habits unknown.

DISCUSSION. The male mouthparts, front leg, wing, and genitalia were illustrated by Hardy (1965, Fig. 156). This species is probably related to *D. acanthostoma* and *D. anoplostoma*, and shares their unusual Hawai'i + Maui Nui distribution. This relationship and the collection localities suggest it is a sap flux breeder on *Acacia koa* in wet forest.

Drosophila peloristoma n. sp.

Fig. 20

DIAGNOSIS. Generally dark-colored, but somewhat more reddish-brown than related species; distinguished from the similar *D. semifuscata* by having the pleura all brown.

DESCRIPTION. ♂. **Head.** Front brown, pollinose; fronto-orbital plates and ocellar triangle less pollinose, shining at some angles. Anterior reclinate seta slightly posterior of the proclinate, about 3/4 as long; posterior reclinate nearly 2.5 times as long as the anterior. Ocellar setae slightly shorter than the posterior reclinate. Vertical setae normal in position, slightly longer than the posterior reclinate. Eyes with short, inconspicuous pile, about as long as one facet. Face brown, with a median carina. Antenna brown; first segment yellowish; arista with 5–6 dorsal and 2 ventral rays in addition to the apical fork. About three strong setae present at the vibrissal angle, slightly longer and stronger than the posterior oral setulae (which are also relatively long), but not strongly differentiated from them. Gena brown; about as wide as 3 eye facets. Palp yellow, broad, flattened, setulose, broadest near the apex which is more or less truncate; with a moderately strong apical seta, but not much longer than the thinner setulae. Labellar sclerite broad, apical margin evenly curved, the two sides often held close together in pinned specimens; margin armed with 4 short ventrally-directed spines (not conspicuous on cursory inspection) along with numerous hair-like setae (Fig. 20A). **Thorax.** Entirely brown and pollinose, except for the humerus which is yellow anteriorly. Two humeral setae, the ventral nearly as long as the dorsal. Acrostichal setulae in about 6–10 irregular rows. Two pairs of dorsocentral setae, the anterior about 3/4 as long as the posterior. Halteres pale brown, stems darker. **Legs.** Yellow, femora slightly tinged with brown. Front tibia lacking ciliation. Front basitarsus with two irregular rows of elongate cilia,

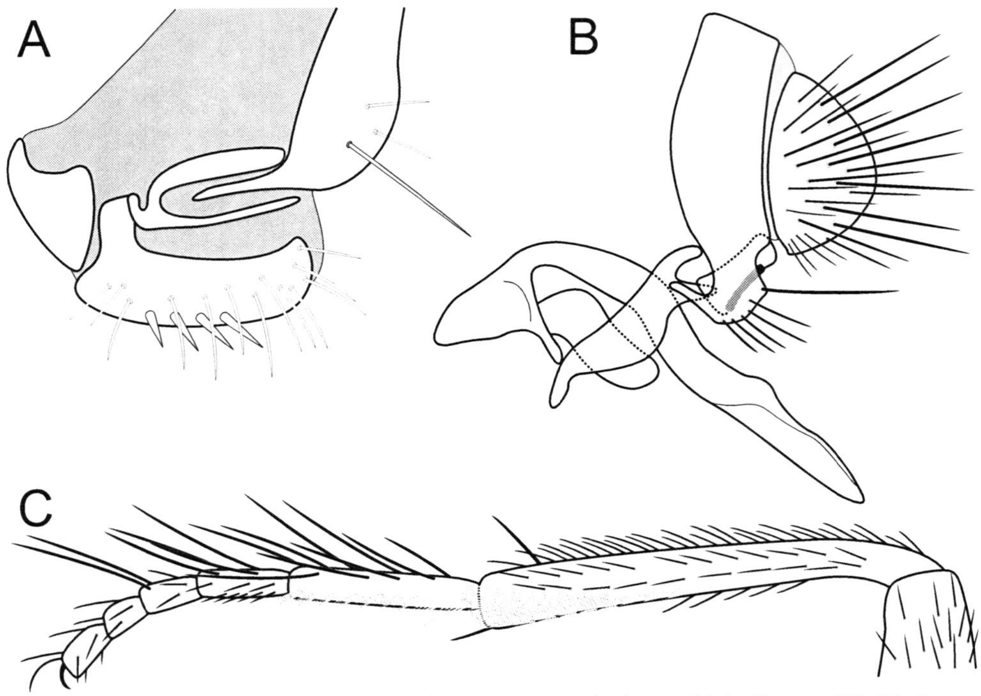

Fig. 20. *Drosophila peloristoma* male. (A) Lateral view of labellum. (B) Right front leg, anterior view.

one dorsal and anterodorsal, with about 3 in the former and 5 in the latter; second segment with two in each row, third segment with one in each; remaining segments without long cilia (Fig. 20C). **Wings.** Faintly infuscated on anterior half, more faintly so around veins M, CuA_1, and the dm–cu crossvein; sometimes hardly visible. Costal fringe long, extending about 3/4 the distance between the apex of R_{2+3} and R_{4+5}. **Abdomen.** Entirely brown. Epandrium slightly wider medially than either dorsally or ventrally; anteroventral spine curled under, thus appearing narrow in lateral view. Aedeagal apodeme about twice as long as high, narrowing toward the apex; ventral margin slightly sinuate, not strongly concave. Aedeagus narrow at the base, broader beyond, strongly bent, basal angle acute, apex elongate; preapical protuberance small, low (Fig. 20B).

♀. Identical to the male with the following exceptions. **Head.** Palp brown, apex broadly rounded. Labellum unmodified. **Legs.** Front legs without elongate cilia or setae. **Abdomen.** Ovipositor brown, broadly triangular, apex pointed.

MEASUREMENTS. N = 4♂. TL = 1.49 (1.25–1.68) mm; WL = 2.77 (2.51–3.00) mm; TL/WL = 0.5 (0.5–0.6); HW = 1.20 (1.10–1.27) mm; HW/FS = 2.2 (2.2–2.3); HW/TL = 0.8 (0.8–0.9); CI = 3.9 (3.6–4.1); 4V = 1.4 (1.3–1.4); 5X = 1.3 (1.2–1.3); 4C = 0.6 (0.5–0.6); M = 0.4. N = 2♀. TL = 1.59 (1.56–1.63) mm; WL = 3.08 (2.97–3.20) mm; TL/WL = 0.5; HW = 1.25 (1.22–1.29) mm; HW/FS = 2.3; HW/TL = 0.8; CI = 3.6 (3.6–3.7); 4V = 1.3 (1.2–1.3); 5X = 1.3; 4C = 0.5 (0.5–0.6); M = 0.4.

TYPES. O'AHU: Holotype ♂ (BPBM 16738) and allotype ♀, Mokulēi'a, ex *Osmanthus* [=*Nestegis*] sap flux, 28.iii.1971, SLM.

PARATYPES. O'AHU: 1♂, Pūpūkea, CH5.1, reared ex koa flux, 1963, WBH. 1♀, Mt. Tantalus, C53, vi.1963, HLC. 4♂, Pālehua, 2200 ft., ex *Acacia koa* flux, 18.iv.1970 [no collector]. 1♂, same data as holotype. 1♂, Pu'u Kaua, 2200 ft., reared ex *Myrsine* flux, Q12, 10.x.1971, SLM (all UHIM).

DISTRIBUTION & ECOLOGY. O'ahu. Breeds in sap flux of various trees.

ETYMOLOGY. From the Greek *peloris* (clam) + *stoma* (mouth), referring to the appearance of the broad, rounded labellar sclerites.

DISCUSSION. The short, stout, ventrally-directed spines on the labellum are distinctive but not particularly conspicuous unless the mouthparts are dissected, and may be missed if one is not specifically looking for them. The spines bear some resemblance to those of *D. lelolua* and may indicate a tenuous relationship with other members of the *semifuscata* subgroup. However, the rearing of this species from sap flux, the wing markings, and its general appearance lead us to place it here. This species is referred to as "Q12" in Magnacca et al. (2008).

Drosophila sadleria Bryan

Drosophila sadleria Bryan, 1938:41. Holotype ♂, O'ahu, Niu, ex rachis of *Sadleria* fern, 11.xii.1910, OHS. BPBM 819 [examined].

DIAGNOSIS. Very close to *D. xuthoptera*, separated by the characters in the key. See Discussion.

DESCRIPTION. ♂. **Head.** Front yellow; fronto-orbital plates and ocellar triangle brown, pollinose. Anterior reclinate seta distinctly posterior of the proclinate (almost halfway to the posterior reclinate), about half as long; posterior reclinate about 2.5 times as long as the anterior. Ocellar setae about as long as the posterior reclinate. Vertical setae normal in position, slightly longer than the posterior reclinate. Eyes with short, inconspicuous pile, about as long as one facet. Face yellow, with a rounded carina dorsomedially. Antenna yellow; arista with 5 dorsal and 2 ventral rays in addition to the apical fork. Several elongate oral setae present at the vibrissal angle (slightly longer than the posterior oral setulae, which are also relatively long), but no distinctly stronger oral vibrissae. Gena yellow, narrowly brown along the oral margin; about as wide as 3 eye facets. Palp yellow, flattened, broadest near the apex, setulose; with an elongate subapical seta. Labellum with 2–3 closely-placed strong, brown, curved, inclinate, round, spine-like setae at the dorsal end, probably followed by several shorter and weaker spines but the latter are not distinct; ventral half with several hair-like setae. **Thorax.** Largely yellow, pollinose. Mesonotum with six longitudinal brown stripes; pleura with two

brown stripes, at the dorsal and ventral margins of the anepisternum. Two humeral setae, the ventral about 2/3 as long as the dorsal. Acrostichal setulae in about 8 irregular rows. Two pairs of dorsocentral setae, the anterior about 3/4 as long as the posterior. **Legs.** Yellow. Front tibia lacking ciliation. Front basitarsus with about 11 elongate dorsal cilia over the apical 3/5 and second segment with about 4 elongate dorsal cilia; remaining segments lacking long cilia. **Wings.** Largely hyaline, with distinct marks over the r–m and dm–cu crossveins and apices of R_{2+3}, R_{4+5}, and M. Costal fringe extending about half the distance between the apex of R_{2+3} and R_{4+5}.
♀. Unknown.

DISTRIBUTION & ECOLOGY. O'ahu; known only from the type. Reared from *Sadleria* ('ama'u, Blechnaceae) fern rachises (Bryan 1938). Breeding habitat apparently similar to that of *D. apicipuncta*, but should be confirmed with more specimens.

DISCUSSION. The male front tarsus and wing were illustrated by Hardy (1965, Fig. 184). This species was previously thought to belong to the *ciliated tarsus* group because Hardy did not mention the spinose mouthparts (and the original description incorrectly stated that the specimen was a female). While all details cannot be seen, it is clear that they are very similar to those of *D. apicipuncta*, *D. bridwelli*, and other species of this subgroup. The type has an additional orbital seta on the left side in between the anterior reclinate and proclinate; presumably this is an aberration, since the right side has the typical chaetotaxy. It is likely that this species differs from *D. xuthoptera* in more significant features than those given in the key, but none can be seen without dissecting the mouthparts and genitalia of the type.

Drosophila semifuscata Hardy

Drosophila semifuscata Hardy, 1965:460–461. Holotype ♂, Maui, Makamaka'ole Valley, vi.1953, DEH. BPBM 6441 [examined].

DIAGNOSIS. Anterior half of wing infuscated; labellum lacking conspicuous spines. Very close to *D. peloristoma*, differing most clearly in having the pleura largely yellow with brown markings, rather than entirely brown.

DESCRIPTION. ♂. **Head.** Front brown, darker and more pollinose on fronto-orbital plates and ocellar triangle. Anterior reclinate seta slightly posterior of the proclinate, about 4/5 as long; posterior reclinate twice as long as the anterior. Ocellar setae about as long as the posterior reclinate. Vertical setae normal in position, slightly longer than the posterior reclinate. Eyes with short, inconspicuous pile, about as long as one facet. Face brown, with a rounded carina dorsomedially. Antenna dark brown, paler on the apicodorsal surface of the second segment; arista with 6 dorsal and 2 ventral rays in addition to the apical fork. Several elongate oral setae present at the vibrissal angle (slightly longer than the posterior oral setulae, which are also relatively long), but no distinctly stronger oral vibrissae. Gena yellow, narrowly brown along the oral margin; about as wide as 2 eye facets. Palp yellow tinged with brown, paler towards the base, flattened, broadest near the apex which is truncate; setulose, with a strong apical and a subapical seta. Labellum fringed with hair-like setae but without conspicuous spines. **Thorax.** Mesonotum entirely brown, pollinose; humerus paler. Pleura generally yellowish, with a distinct brown stripe along dorsal margin of anepisternum, an oval brown spot covering bases of the katepisternal setae, and broadly brown over the anepimeron. Two humeral setae, the ventral slightly shorter than the dorsal. Acrostichal setulae in about 8 irregular rows. Two pairs of dorsocentral setae, the anterior about 2/3 as long as the posterior. **Legs.** Yellow. Front tibia lacking ciliation. Front basitarsus with about 10 elongate cilia in three irregular dorsal to anterodorsal rows; second segment with 5 cilia continuing these rows; third segment with a pair of cilia near the apex; remaining segments lacking long cilia. **Wings.** Entirely infuscated anterior of R_{4+5}, as well as over the dm–cu crossvein and basally along M to the crossvein. Costal fringe extending about 3/5 the distance between the apex of R_{2+3} and R_{4+5}. **Abdomen.** Each segment yellowish brown anteriorly, dark brown on posterior half.

♀. Identical to the male with the following exceptions. **Head.** Palp with only one strong subapical seta. Two strong oral vibrissae present. **Thorax.** Pleura entirely brown. **Legs.** Femora brownish. Front legs without cilia. **Abdomen.** Entirely dark brown. Ovipositor broadly triangular, apex acute.

MATERIAL EXAMINED. MAUI: 1♀, allotype, Waikapu Valley, 23.iii.1924, OHS (BPBM).

DISTRIBUTION & ECOLOGY. Maui; known only from the holotype and allotype. Breeding habits unknown; probably breeds in sap fluxes like *D. peloristoma* and other members of the subgroup.

DISCUSSION. The male front leg and wing were illustrated by Hardy (1965, Fig. 186c–d). The mouthparts have never been dissected, and it is possible that, like *D. peloristoma*, *D. semifuscata* possesses small labellar spines that are not visible on the intact specimen. The female resembles *D. anoplostoma*, which was unknown at the time *D. semifuscata* was described. The male of *D. semifuscata* is considerably different in coloration, and their collections were widely separated in both time and space. However, such differences are not unprecedented; therefore, until more specimens are available we are leaving the identification as is.

Drosophila wawae **n. sp.**
Fig. 21

DIAGNOSIS. Distinguished from other species in the pale-bodied complex by the relatively short dorsal labellar spines (about as long as the other spines, rather than significantly longer as in other species; compare Figs. 19A and 21A) and by having long cilia over the first three segments of the tarsus.

DESCRIPTION. ♂. **Head.** Front brown, darker and more pollinose on fronto-orbital plates and ocellar triangle. Anterior reclinate seta slightly posterior of the proclinate, about half as long; posterior reclinate about 2.5 times as long as the anterior. Ocellar setae about as long as the posterior reclinate. Vertical setae normal in position, slightly longer than the posterior reclinate. Eyes with short, inconspicuous pile, about as long as one facet. Face largely yellow, with a rounded carina dorsomedially; ventrolateral corners brown. Antenna yellow, third segment brown; arista with 6 dorsal and 3 ventral rays in addition to the apical fork. Several elongate oral setae present at the vibrissal angle (slightly longer than the posterior oral setulae, which are also relatively long), but no distinctly stronger oral vibrissae. Gena yellow, narrowly brown along the oral margin; about as wide as 4–5 eye facets. Palp yellow, flattened, broadest near the apex but only weakly clavate, setulose; with an elongate subapical seta. Labellum with a cluster of 5 strong, brown, round, spine-like setae at the dorsal end, the first two closely appressed and appearing to be one, only slightly longer than the others; followed ventrally by about 10 hair-like setae. (Fig. 21A) **Thorax.** Mesonotum pale yellowish brown to rufous, pollinose, with a faint, narrow, brown stripe medially and one down each dorsocentral row. Pleura largely yellow, with a narrow brown stripe along the dorsal margin of the anepisternum and anepimeron, and a broader one along the ventral margin. Two humeral setae, the ventral nearly as long as the dorsal. Acrostichal setulae in about 8 irregular rows. Two pairs of dorsocentral setae, the anterior about 2/3 as long as the posterior. Halteres yellow, stems darker. **Legs.** Brownish yellow. Front tibia lacking ciliation. Front basitarsus with a total of about 15 elongate cilia in irregular dorsal and anterodorsal rows; second segment with about 8 and third with about 4, remaining segments without long cilia (Fig. 21D). **Wings.** Infuscated along the entire anterior margin and around the apex to vein M, sometimes connecting with a large infuscation over the dm–cu crossvein; margins indistinct, diffuse. Costal fringe extending about 1/3 the distance between the apex of R_{2+3} and R_{4+5}. **Abdomen.** Largely dark brown, each segment yellow anterolaterally. Epandrium parallel-sided; anteroventral spine curled under, thus appearing narrow in lateral view. Aedeagal apodeme about twice as long as high, strongly concave ventrally (Fig. 21B). Aedeagus narrow at the base, broader beyond, strongly bent, basal angle acute; apex broken off in dissected specimen. Surstyli with elongate medially-directed setae (Fig. 21C).
♀. Unknown.

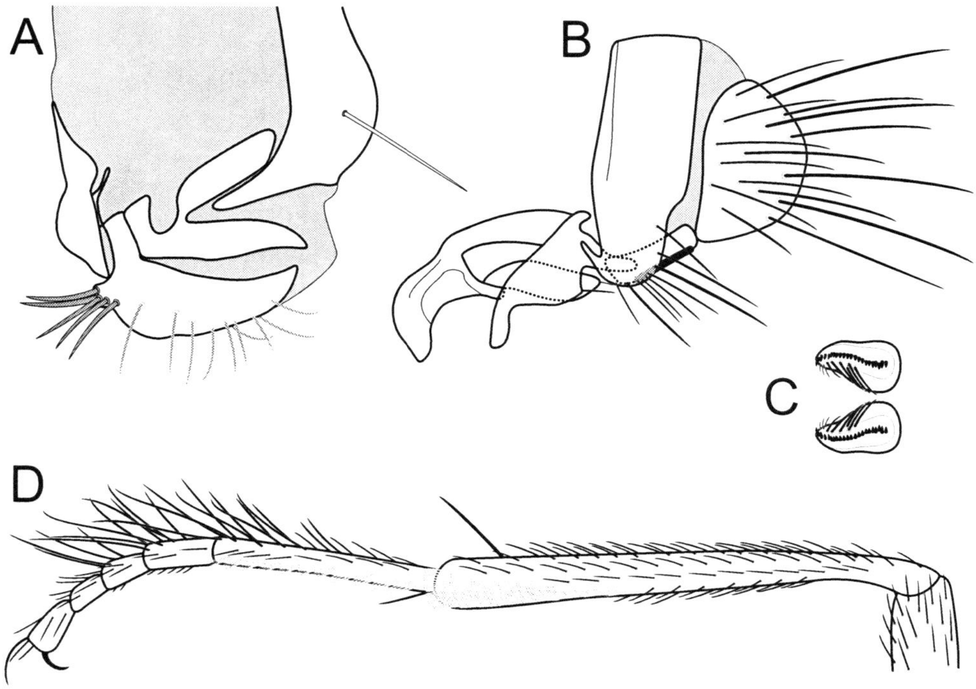

Fig. 21. *Drosophila wawae* male. (A) Lateral view of labellum. (B) Terminalia (apex of aedeagus missing). (C) Surstyli, ventral view. (D) Right front leg, anterior view.

MEASUREMENTS. N = 1♂. TL = 2.18 mm; WL = 4.92 mm; TL/WL = 0.4; HW = 1.59 mm; HW/FS = 2.1; HW/TL = 0.7; CI = 4.4; 4V = 1.2; 5X = 1; 4C = 0.5; M = 0.3.

TYPES. HAWAI'I: Holotype ♂ (BPBM 16739), Kehena Ditch Trail, S. Kohala Mts., N17, 22–24.vii.1969, WBH.

PARATYPES. 1♂, Kūlani Cone, 5100 ft., K64, 2.x.1967, HLC (UHIM).

DISTRIBUTION & ECOLOGY. Hawai'i; rare. Breeding habits unknown.

ETYMOLOGY. From the Hawaiian *wāwae*, paw or foot, referring to the resemblance of the mouthparts to feline claws.

Drosophila xuthoptera Hardy

Drosophila xuthoptera Hardy, 1965:516–519. Holotype ♂, Moloka‘i, Pu‘u o Ka‘eha, vii.1953, DEH. BPBM 6472 [examined].

DIAGNOSIS. Groups with the *D. apicipuncta* complex based on the long dorsal spines of the labellum, and generally yellow-brown thorax; separated from most related species by the combination of elongate cilia on only the first two segments of the front tarsus (including the apical half of the basitarsus), and usually having the anterior wing margin infuscated. Almost identical to *D. sadleria* in body coloration and tarsal ciliation, but separated by the details of setation given in the key.

DESCRIPTION. ♂. **Head.** Front brown above, yellow near frontal suture; pollinose. Anterior reclinate seta slightly posterior of the proclinate, about 2/3 as long; posterior reclinate about 3 times as long as the anterior. Ocellar setae about as long as the posterior reclinate. Vertical setae normal in position, about as long as the posterior reclinate. Eyes with short, inconspicuous pile, about as long as one facet. Face yellow, with a rounded median carina. Antenna yellow, tinged with brown on segment 3; arista with 5–6 dorsal and 2 ventral rays in addition to the apical fork. Several moderately long setae present at the vibrissal angle, not differentiated from the other oral setulae, which are also relatively long. Gena yellow, narrowly brown along the oral margin; about as wide as 4 eye facets. Palp yellow, flattened, setulose, broadest before the apex which is rounded; with a single strong apical seta. Labellar sclerite with 2 closely-placed strong, brown, curved, inclinate, round, spine-like setae at the dorsal end (often appearing as one), followed by 2–3 shorter and weaker spines; ventral half with several hair-like setae. **Thorax.** Yellow, tinged faintly with brown. Pleura with a dark brown stripe at the dorsal margin of the anepisternum; remainder of anepisternum and anepimeron variable, ranging from entirely yellow, to yellow only medially on the anepisternum, to almost entirely brown; posterodorsal margin of katepisternum also brown in dark individuals. Two humeral setae, the ventral nearly as long as the dorsal. Acrostichal setulae in about 6–10 irregular rows. Two pairs of strong dorsocentral setae, the anterior about 3/4–4/5 as long as the posterior; 3–4 additional short setae, about 1/3 as long as the anterior dorsocentral and twice as long as the other setulae, are present in the dorsocentral row, including one presutural. Halteres yellow, stems brown. **Legs.** Yellow. Front tibia lacking ciliation. Front basitarsus with about 8 elongate cilia over the apical 2/3 in an irregular anterodorsal row and 2 dorsal cilia on the apical 1/6; second segment with about 6 cilia scattered over the dorsal surface; remaining segments without long cilia. **Wings.** Vein M between the crossveins, dm–cu crossvein, and entire anterior margin and apex of wing infuscated; markings faint and diffuse in pale-bodied specimens, especially the anterior margin. Costal fringe extending about 1/3 the distance between the apex of R_{2+3} and R_{4+5}. **Abdomen.** Each segment dark brown posteriorly, with a yellow anterior band.

♀. Identical to the male with the following exceptions. **Head.** Two strong oral vibrissae present; other oral setae also longer and stronger. Labellum unmodified. **Legs.** Front legs without cilia. **Abdomen.** Ovipositor broadly triangular, apex blunt.

MATERIAL EXAMINED. MOLOKA'I: 1♀ paratype, Pu'u Ali'i, 4200 ft., vii.1953, DEH. 1♂ 1♀, Hanalilolilo, 3000 ft., C139.5, 3.viii.1965, HLC. 1♂, Pu'u Kolekole, M51, 8–10.viii.1969, HLC (all UHIM). MAUI: 1♂, Palikū, 6500 ft., vi.1953, C. R. Joyce. 1♂, Waikamoi, 4000 ft., viii.1958, DEH. 1♂ 1♀, Palikū, 6500 ft., CH1.28, 23.vii.1963, WBH. 3♂, Palikū, 27.vii.1964, H. Takada. 1♂, Waikamoi, C104.26, 8.vii.1964, HLC. 1♂, Waikamoi, 4300 ft., C104, 9.vii.1964, HLC. 1♂, Waikamoi, 21.x.1964, DEH. 1♂, Waikamoi, C140.15, 12.viii.1965, HLC. 1♂, 'Ukulele Camp Trail, 27.i.1967, KYK. 1♂, Hana'ula, 9–10.vii.1968, KYK (all above at UHIM). 1♂, Palikū, crater wall, 6600 ft., 1.viii.2007, KNM (UCPO). 1♂, Palikū, crater wall, 6400 ft., 2.viii.2007, KNM (UCPO). 1♂ 1♀, Ha'ikū Uka, Heed Trail, 4200 ft., sweeping vegetation and ground, 6.viii.2007, KNM (UCPO).

DISTRIBUTION & ECOLOGY. Maui, Moloka'i. Breeding habits unknown.

DISCUSSION. The male front tarsus, wing, and genitalia, and ovipositor were illustrated by Hardy (1965, Fig. 214). This is the most common and also the most variable member of this subgroup. Specimens from the Waikamoi area of East Maui tend to have a paler thorax and weaker wing marks than others, but the differences are not absolute and no structural characters can be found to separate them. Also see Discussion under *D. sadleria*.

Drosophila z-notata Bryan

Drosophila z-notata Bryan, 1934:437. Holotype ♀, O'ahu, Waiāhole, 28.iii.1915, OHS. BPBM 812 [examined].

DIAGNOSIS. Both sexes immediately distinguishable from all other species by the distinctive wing marks, infuscated around the entire margin to CuA_1, with stripes along all veins except basal segment of CuA_1 and an S-shaped hyaline area near the apex. Male mouthparts similar to those of the *D. apicipuncta* complex.

DESCRIPTION. ♂. **Head.** Front yellowish brown, pollinose; darker and more pollinose on fronto-orbital plates and ocellar triangle. Anterior reclinate seta slightly posterior of the proclinate, nearly as long but distinctly weaker; posterior reclinate about twice as long as the anterior. Ocellar setae about as long as the posterior reclinate. Vertical setae normal in position; outer about as long as the posterior reclinate, inner slightly longer. Eyes with short, inconspicuous pile, about as long as one facet. Face brown, with a rounded median carina. Antenna yellow-brown; arista with 6 dorsal and 3 ventral rays in addition to the apical fork. No

oral vibrissae or other elongate oral setae present. Gena yellow, narrowly brown along the oral margin; about as wide as 3 eye facets. Palp yellow-brown, flattened, sparsely setulose, almost parallel-sided; with a strong apical and a smaller subapical seta. Labellar sclerite with 2 closely-placed strong, brown, curved, inclinate, round, spine-like setae at the dorsal end (often appearing as one), followed by 2–3 shorter and weaker spines, which grade into hair-like setae. **Thorax.** Brown, mesonotum with 3 narrow gray longitudinal stripes, one medial and one just lateral of each dorsocentral row; pleura with a darker stripe along dorsal margin of anepisternum and a yellow stripe along the ventral margin. Two humeral setae, the ventral about 2/3 as long as the dorsal. Acrostichal setulae in about 6–8 irregular rows. Two pairs of strong dorsocentral setae, the anterior about 3/4 as long as the posterior. Halteres brown. **Legs.** Yellow. Front tibia lacking ciliation. Front basitarsus with 6 elongate cilia over the apical half in two irregular dorsal rows; second segment with 2–3 dorsal cilia; remaining segments without long cilia. **Wings.** Largely infuscated, including the dm–cu crossvein, entire anterior half, and apical margin around to CuA_1. Most of vein R_{4+5} beyond the r–m crossvein infuscated, with a hyaline interruption just anterior of the dm–cu crossvein. Region of wing posterior and basal of crossveins hyaline. Costal fringe extending about 1/3 the distance between the apex of R_{2+3} and R_{4+5}. **Abdomen.** Abdomen of only male specimen is in poor condition and discolored.
♀. Identical to the male with the following exceptions. **Head.** Labellum unmodified. **Thorax.** Mesontal stripes yellowish. **Legs.** Front legs without cilia. **Wings.** Middle segment of vein M between the crossveins also infuscated. **Abdomen.** Entirely dark brown. Ovipositor broadly triangular, apex blunt.

MATERIAL EXAMINED. O'AHU: 2♀ paratypes, same data as holotype (BPBM). 1♂, Punalu'u, OHS [no other data] (UHIM).

DISTRIBUTION & ECOLOGY. O'ahu. Breeding habits unknown.

DISCUSSION. The male head, front tarsus, and genitalia, and female wing and ovipositor were illustrated by Hardy (1965, Fig. 215). The true colors of the body are difficult to discern due to the age of the specimens. Evenhuis (1982) describes the three Waiāhole specimens (which are mounted on one piece of wood) as syntypes, and states that the word "type" pointing to one of them is not in Bryan's handwriting. However, since the original description refers to "type and two paratypes from Waiahole," (Bryan 1934) we consider the one designated on the label to be the holotype regardless of whether Bryan himself wrote it.

setiger subgroup

The four members of this subgroup are quite distinctive from all other Hawaiian *Drosophila*, being shining black on the thorax and head, but extremely similar to one another. The ciliation of the front legs easily separates the species, but the unusual mouthparts are almost identical among them and there are no other structural differences aside from minor variations in setation. Like many other *modified mouthparts* species, they appear to be widespread but are relatively rarely collected.

1. Front basitarsus without unusually long or strong setae, the longest shorter than the segment .... 2.
— Front basitarsus with two long, strong setae at the base, at least as long as the first two tarsal segments combined .... 3.

2. (1) Front tibia lacking long cilia. Kaua'i .... *D eurypeza* Hardy
— Front tibia with a row of 4 anterior cilia over the apical half. O'ahu .... *D. imitator* Hardy

3. (1) Front tibia lacking long cilia or setae (a single short anterior cilium present near the preapical dorsal seta) (Fig. 22C). Hawai'i .... *D. desallei* n. sp.
— Front tibia with a long seta near the middle, almost as long as the tibia. Moloka'i, Lāna'i .... *D. setiger* Grimshaw

Drosophila desallei n. sp.
Fig. 22

Diagnosis. Easily separated from other species in this distinctive subgroup by having very long cilia on the front tarsus but not the tibia.

Description. ♂. **Head.** Front dark brown to black, shining. Anterior reclinate seta about even with the proclinate and smaller, about half as long; posterior reclinate very strong and about 3–4 times as long as proclinate. Ocellar setae about 3/4 as long as the posterior reclinate. Vertical setae normal in position; inner vertical elongate, about 1.3 times as long as the posterior reclinate, outer vertical slightly shorter than posterior reclinate. Eyes with short, inconspicuous pile, about as long as one facet. Face with a rounded carina medially; carina and ventral half white, brown below the antennae. Antenna black; arista with 6–7 dorsal and 2 ventral rays in addition to the apical fork. No strong oral vibrissae. Gena tan, with a dark brown stripe along the oral margin. Palp dark brown, flattened, clavate, broadest at about 2/3 the length, sparsely setulose; one subapical seta present. Labellum with a sclerotized anterior appendage at the dorsal end; this is about half as long as the apex of the labellum and has 5 apical setae (Fig. 22A). Concealed behind the

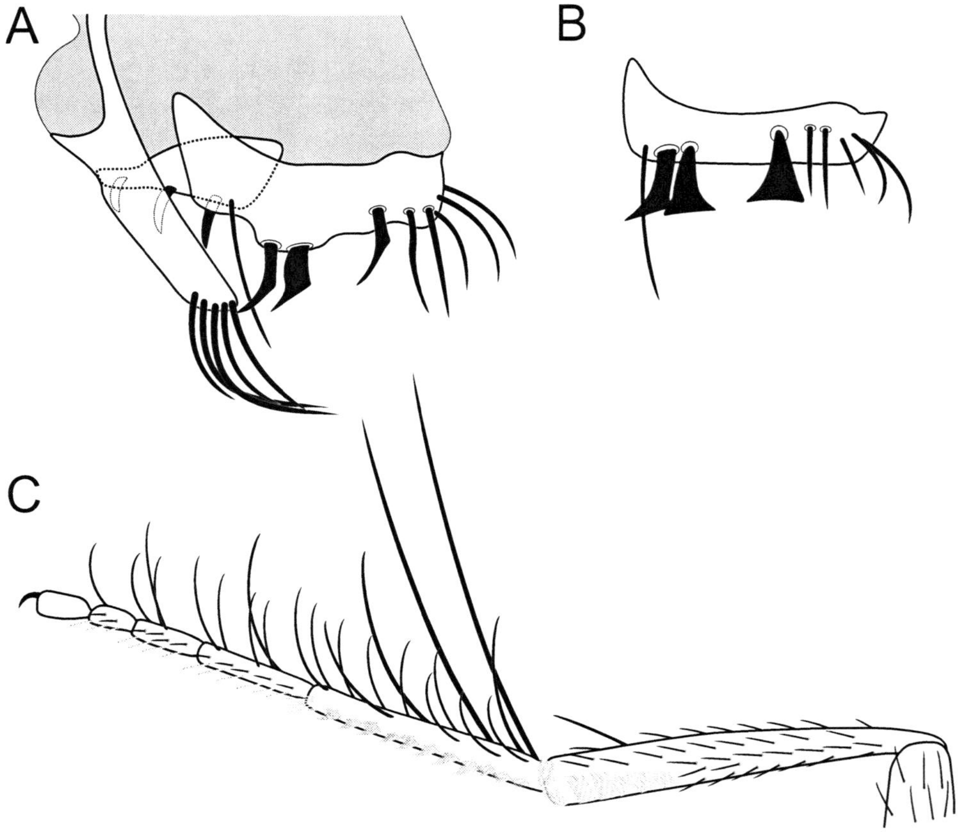

Fig. 22. *Drosophila desallei* male. (A) Lateral view of labellum. (B) Oblique posteroventral view of labellar sclerite, showing the shape of the black spines. (C) Right front leg, anterior view.

appendage are three brown, spine-like, declinate setae. Ventral of the appendage are three black, triangular spines (broadest at the apex); the first two are close together, while the third is close to two brown ventral setae, similar to but weaker than those near the appendage (Fig. 22B). **Thorax.** Entire thorax dark brown to black, shining. One humeral seta; one setula is longer and probably the ventral humeral, but it is very weak and only about 1/3 as long as the strong dorsal humeral. Acrostichal setulae in 6 rows. Two pairs of dorsocentral setae, the anterior nearly as long as the posterior. Halteres brown, stems somewhat paler. **Legs.** Coxae and femora yellow, tinged with brown, varying in the three specimens from dark brown to entirely yellow; tibiae and tarsi always yellow. Front tibia with only a single anterior cilia near the apex, below the preapical dorsal seta. Front basitarsus with an anterior row of 5 and a dorsal row of 6 long, curved cilia extending the entire length; both continue with two and one of each on the second and third segments respectively (Fig. 22C). Three much longer posterodorsal cilia are also present on the basal half of the basitarsus; the basal two are about as long as the first two tarsal segments combined, while the third is about as long as the basitarsus. Another posterodorsal is present on the second segment but it is smaller, closer in size to the other cilia. **Wings.** Mostly hyaline, infuscated at the base between the humeral crossvein and R_1, basally extending posteriorly as far as the anal cell. Costal fringe extending 1/4–1/3 the distance between the apex of R_{2+3} and R_{4+5}. **Abdomen.** Entirely dark brown to black and shining, including terminal segments. Genitalia not dissected.

♀. Identical to the male with the following exceptions. **Head.** Posterior reclinate only about twice as long as the proclinate. Face entirely brown. Palp with a long apical seta, nearly as long as the palp. Labellum unmodified. **Legs.** Front legs without long cilia. **Wings.** Infuscation extended slightly beyond R_1. Slightly infuscated anteroapically, but without a distinct mark. **Abdomen.** Ovipositor broadly triangular, apex blunt.

MEASUREMENTS. N = 1♂. TL = 0.97 mm; WL = 2.46 mm; TL/WL = 0.4; HW = 0.79 mm; HW/FS = 2.3; HW/TL = 0.8; CI = 4.4; 4V = 1.3; 5X = 1.4; 4C = 0.5; M = 0.4. N = 1♀. TL = 1.40 mm; WL = 3.02 mm; TL/WL = 0.5; HW = 0.97 mm; HW/FS = 2.2; HW/TL = 0.7; CI = 4.8; 4V = 1.2; 5X = 1.2; 4C = 0.5; M = 0.4.

TYPES. HAWAI'I: Holotype ♂ (BPBM 16715) and allotype ♀, Kawaihae Uka, 5100 ft., 5.x.2006, KNM.

PARATYPE. HAWAI'I: 1♂, Saddle Rd., Kipuka #3, 5200 ft., 9.vii.1967, WBH.

OTHER MATERIAL. HAWAI'I: 1♂, Bird Park, Kīlauea, N26, reared ex *Pisonia* stems, 16.i.1970, WBH.

DISTRIBUTION & ECOLOGY. Hawai'i; known from only a few collections, but probably widespread in wet and mesic forest. Reared from *Pisonia brunoniana* (pāpala kēpau, Nyctaginaceae).

ETYMOLOGY. Named for Rob DeSalle of the American Museum of Natural History, for his contributions to Hawaiian *Drosophila* phylogenetics.

DISCUSSION. The specimen from Bird Park (Kīpuka Puaulu) is clearly a member of the *setiger* subgroup based on the mouthparts and is assumed to be of this species (there are no other representatives known from Hawai'i), but is not included as a paratype because the front legs are damaged.

Drosophila eurypeza Hardy

Drosophila eurypeza Hardy, 1965:266–267. Holotype ♂, Kaua'i, Kōke'e, 3600 ft., vii.1952, DEH. BPBM 6354 [examined].

Diagnosis. See Description.

Description. Identical to *D. desallei* with the following exceptions. **Head.** Posterior reclinate about 2.5–3 times as long as proclinate. Ocellar setae about as long as the posterior reclinate. Both vertical setae about as long as the posterior reclinate. Arista with 7–8 dorsal and 2–3 ventral rays in addition to the apical fork. Triangular spines of male labellum approximately evenly spaced; setae of labellar appendage coherent, appearing as one. **Thorax.** Acrostichal setulae in 8–10 irregular rows. Anterior dorsocentral seta about 2/3 as long as the posterior. **Legs.** Coxae and femora dark brown. Male front tibia without cilia. Male front tarsus with two irregular posterodorsal rows of curved cilia extending the entire length, these 2–3 times as long as the width of the segment. **Wings.** Infuscation extending across the anterior margin to the apex of R_{2+3}; darkest at the base, fading beyond the apex of R_1; female similar to male, without anteroapical infuscation.

Material Examined. Kaua'i: 1♀, allotype, Wai'alae Stream, 3600 ft., viii.1953, DEH (BPBM). 1♂ 1♀ paratypes, Halemanu Swamp, viii.1953, DEH. 1♀ paratype, Alaka'i Swamp, 4000 ft., viii.1953, DEH. 1♂, Kōke'e, Kumuwela Trail, 30.vii.1963, DEH. 1♂, Kōke'e, 22.v.1979, DEH. 1♂, Y64 [no other data], KYK & WDP (all above except allotype at UHIM). 1♀, Pihea Trail, 4100 ft., 21.v.2007, KNM (UCPO). 1♂, Pu'u O Kila Rd., 4050 ft., 21.v.2007, KNM (UCPO).

Distribution & Ecology. Kaua'i, wet forest. Breeding habits unknown.

Discussion. By its setation, *D. eurypeza* is the most plesiomorphic of the *setiger* subgroup species: the inner vertical is not elongated, there are no setae or cilia on the front tibia, and the acrostichal setulae are in the 8–10 irregular rows typical of most *modified mouthparts* species, rather than the 6 relatively regular rows of the other species.

Drosophila imitator Hardy

Drosophila imitator Hardy, 1965:312–314. Holotype ♂, O'ahu, Pūpūkea, xii.1952, DEH. BPBM 6374 [examined].

Diagnosis. See Description.

Description. Identical to *D. desallei* with the following exceptions. **Head.** Inner vertical seta only slightly longer than posterior reclinate. Palp with both apical and subapical setae. **Legs.** Entirely yellow. Male front tibia with 4 long, anterior, apically-directed cilia, these becoming shorter towards the apex such that they all end at nearly the same point, about 1/3 the distance down the basitarsus. Male front basitarsus with 7–8 thin cilia each in irregular anterodorsal and dorsal rows along the full length, 3–4 stronger posterodorsal cilia, and a strong anterodorsal seta at midlength; second segment with 2–3 cilia in each of the three rows; third segment with one in each row. **Wings.** Infuscation extending at least partway along the anterior margin into cell r_1, but much fainter than in cell c; female similar to male, without anteroapical infuscation.

MATERIAL EXAMINED. O'AHU: 1♂, Castle Trail, 11.iv.1970, ex rotten stem *Touchardia*. 1♀, Kōnāhuanui, 3000 ft., 10.v.1981, SLM. 1♀, Kōnāhuanui, 3000 ft., 28.v.1983, WDP (all UHIM).

DISTRIBUTION & ECOLOGY. O'ahu, wet forest; rare. Reared from bark of *Touchardia latifolia* (olonā, Urticaceae).

Drosophila setiger Grimshaw

Drosophila setiger Grimshaw, 1901:64. Type ♂ and ♀, Moloka'i, forest above Pelekunu, 24.viii.1893, RCLP, BMNH [examined].

Drosophila apoxyloma Hardy, 1965:156–157. Holotype ♂, Moloka'i, Pēpē'ōpae, 4000 ft., vii.1959, D. H. Habeck. BPBM 6297 [examined]. **n. syn.**

DIAGNOSIS. See Description.

DESCRIPTION. Identical to *D. desallei* with the following exceptions. **Head.** Anterior reclinate seta slightly posterior of proclinate. **Legs.** Entirely yellow. Male front tibia with a single anterior seta about halfway down; very long, about 4/5 as long as the tibia. Male front basitarsus with 2 closely placed, strong dorsal setae at the base, about as long as the first three tarsal segments combined; also with 6–7 thin cilia each in irregular anterodorsal and dorsal rows along the full length, and 2 posterodorsal cilia on the apical third; second segment with 2 cilia in each of the three rows; third and fourth segments both with one in each row. **Wings.** Infuscation extending at least partway along the anterior margin, sometimes to the apex of cell r_1, but much fainter than in cell c; female similar to male, without anteroapical infuscation.

MATERIAL EXAMINED. Identified as *D. setiger*: MOLOKA'I: 1♂, Pu'u Ali'i, 4200 ft., vii.1953, DEH [labelled "homotype" and "cf type"]. MAUI: 2♀, Waikamoi, 4000 ft., 19.vii.1965, JWB. Identified as *D. apoxyloma*: MOLOKA'I: 1♀, Pēpē'ōpae, 4000 ft., 30.vii.1959, DEH. LĀNA'I: 1♂, Lāna'ihale, 3300 ft., 25.iii.1966, JWB (all UHIM).

DISTRIBUTION & ECOLOGY. Moloka'i and Lāna'i, wet forest; possibly Maui, but only females collected there. Breeding habits unknown.

DISCUSSION. We have examined the types of *D. setiger*, a specimen of *D. setiger* labelled "homotype" by Hardy, the holotype of *D. apoxyloma*, and an additional male specimen in the UHIM identified as *D. apoxyloma*. In all of these, the male front leg setation is identical, as is shown in Hardy's illustrations (Hardy, 1965; Figs. 35d and 188b). The only character separating the two is the greater wing infuscation of *D. apoxyloma* mentioned by Hardy (1965), extending across the costal margin rather than being restricted to cell c. However, it is apparent even from the limited number of specimens available that this character is highly variable in the subgroup, and no distinction can be made between specimens with a greater or lesser degree of infuscation. In fact, the type of *D. setiger* has cell r_1 infuscated, albeit faintly, nearly to the apex. The two are also found in the same region of montane wet forest on Moloka'i. Therefore, we reduce *D. apoxyloma* to a new junior synonym of *D. setiger*.

LITERATURE CITED

Bonacum, J. 2001. Molecular systematics of the Hawaiian Drosophilidae. Ph.D. thesis, Yale University, New Haven.

Bryan, E. H. 1934. A review of the Hawaiian Diptera, with descriptions of new species. Proceedings of the Hawaiian Entomological Society 8:399–468.

Bryan, E. H. 1938. Key to the Hawaiian Drosophilidae and descriptions of new species. Proceedings of the Hawaiian Entomological Society 10:25–42.

Evenhuis, N. L. 1982. Catalog of entomological types in the Bishop Museum. Diptera: Drosophilidae. Pacific Insects 24:318–367.

Grimshaw, P. H. 1901. Diptera. Fauna Hawaiiensis 3(1):1–77.

Hardy, D. E. 1965. Insects of Hawaii vol. 12. Diptera: Cyclorrhapha II, Series Schizophora, Section Acalypterae I. Family Drosophilidae. University of Hawaii Press, Honolulu. 814 pp.

Hardy, D. E. 1966. Descriptions and notes on Hawaiian Drosophilidae (Diptera). University of Texas Publications 6615:195–244.

Hardy, D. E., and Kaneshiro, K. Y. 1968. New picture-winged *Drosophila* from Hawaii. University of Texas Publications 6816:171–262.

Hardy, D. E., and Kaneshiro, K. Y. 1969. Descriptions of new Hawaiian *Drosophila*. University of Texas Publications 6918:39–54.

Hardy, D. E., and Kaneshiro, K. Y. 1975a. Studies in Hawaiian *Drosophila*, miscellaneous new species, no. 1. Proceedings of the Hawaiian Entomological Society 22:57–64.

Hardy, D. E., and Kaneshiro, K. Y. 1975b. Studies in Hawaiian *Drosophila*, modified mouthparts species no. 1: mitchelli subgroup. Proceedings of the Hawaiian Entomological Society 22:51–55.

Heed, W. B. 1968. Ecology of the Hawaiian Drosophilidae. University of Texas Publications 6816:387–419.

Kaneshiro, K. Y. 1997. R.C.L. Perkins' legacy to evolutionary research on Hawaiian Drosophilidae (Diptera). Pacific Science 51:450–461.

Magnacca, K. N., Foote, D., and O'Grady, P. M. 2008. A review of the endemic Hawaiian Drosophilidae and their host plants. Zootaxa 1728:1–58.

Magnacca, K. N., and O'Grady, P. M. 2007. A subgroup structure for the *modified mouthparts* species group of Hawaiian *Drosophila*. Proceedings of the Hawaiian Entomological Society 38:87–101.

Magnacca, K. N., and O'Grady, P. M. 2008. Revision of the 'nudidrosophila' and 'ateledrosophila' species groups of Hawaiian *Drosophila* (Diptera: Drosophilidae), with descriptions of twenty-two new species. Systematic Entomology 33:395–428.

O'Grady, P. M., Kam, M. W. Y., Val, F. C., and Perreira, W. D. 2003. Revision of the *Drosophila mimica* subgroup, with descriptions of ten new species. Annals of the Entomological Society of America 96:12–38.

Spieth, H. T. 1966. Courtship behavior of endemic Hawaiian *Drosophila*. University of Texas Publications 6615:245–313.

Takada, H., and Yoon, J. S. 1989. Three new *Drosophila* species (Diptera: Drosophilidae) from British Columbia, Hawaii, and the Canary Islands. Entomological News 100:111–121.

Lightning Source Inc.
LaVergne, TN USA
14 August 2009

154803LV00004B/5/P

9 780520 098732